ANCIENT METEOROLOGY

In antiquity meteorology included the study of the weather, and also the study of geological, seismological and astronomical phenomena, such as earthquakes and comets. *Ancient Meteorology* explores Greek and Roman approaches and attitudes to this broad subject.

Given the importance of farming in ancient society, it is not surprising that so much was written about the prediction and explanation of the weather and about how to respond to its cruelties and to its kindnesses. But meteorology was not just for farmers, and it was not just a practical matter. Poets, philosophers and physicians were also interested, and used the subject to raise important questions about the nature of the world and how we understand it, about the unity and character of the cosmos, and about the relationship between meteorology and the divine.

Liba Taub discusses the variety of ancient texts which communicate meteorological and scientific ideas, from Homeric epic and the didactic poetry of Hesiod, Aratus and Lucretius, to works such as Aristotle's *Meteorology*, the Hippocratic medical treatise on *Airs, Waters, Places* and Seneca's *Natural Questions*. The range and diversity of this literature highlights questions of intellectual authority in antiquity and illustrates the lively engagement of ancient authors with the work of their predecessors.

Ancient Meteorology will be a valuable and stimulating resource for classicists and readers interested in the history of science.

Liba Taub is Director and Curator of the Whipple Museum of the History of Science, in the Department of History and Philosophy of Science at the University of Cambridge, and a Fellow of Newnham College. She is the author of *Ptolemy's Universe: The Natural and Ethical Foundations of Ptolemy's Astronomy*.

SCIENCES OF ANTIQUITY
Series Editor: Roger French
*Director, Wellcome Unit for the History of Medicine,
University of Cambridge*

Sciences of Antiquity is a series designed to cover the subject-matter of what we call science. The volumes discuss how the ancients saw, interpreted and handled the natural world, from the elements to the most complex of living things. Their discussions on these matters formed a resource for those who later worked on the same topics, including scientists. The intention of this series is to show what it was in the aims, expectations, problems and circumstances of the ancient writers that formed the nature of what they wrote. A consequent purpose is to provide historians with an understanding of the materials out of which later writers, rather than passively receiving and transmitting ancient 'ideas', constructed their own world-view.

ANCIENT ASTROLOGY
Tamsyn Barton

ANCIENT NATURAL HISTORY
Histories of nature
Roger French

COSMOLOGY IN ANTIQUITY
M.R. Wright

ANCIENT MATHEMATICS
S. Cuomo

ANCIENT METEOROLOGY
Liba Taub

TO MY PARENTS

ANCIENT METEOROLOGY

Liba Taub

LONDON AND NEW YORK

First published 2003
by Routledge
11 New Fetter Lane, London EC4P 4EE

Simultaneously published in the USA and Canada
by Routledge
29 West 35th Street, New York, NY 10001

Routledge is an imprint of the Taylor & Francis Group

© 2003 Liba Taub

Typeset in Garamond by
Florence Production Ltd, Stoodleigh, Devon
Printed and bound in Great Britain by
TJ International, Padstow, Cornwall

British Library Cataloguing in Publication Data
A catalogue record for this book is available from the British Library

Library of Congress Cataloging in Publication Data
Taub, Liba Chaia, 1954–
Ancient meteorology/Liba Taub.
p. cm. – (Sciences of antiquity)
Includes bibliographical references and index.
1. Meteorology – Greece – History.   2. Meteorology – Rome – History.
3. Science, Ancient.   I. Title.   II. Series.
QC985.5.G8T38 2003
551.5′0938 – dc21   2002037044

ISBN 0–415–16195–9 (hbk)
ISBN 0–415–16196–7 (pbk)

CONTENTS

ILLUSTRATIONS

A NOTE ON THE SPELLING OF GREEK NAMES AND TERMS

For the most part (but not always), I have adopted a 'latinized' spelling of Greek names and terms (e.g. Eudoxus, rather than Eudoxos, Callippus, rather than Kallippos), to conform to general usage.

ABBREVIATIONS

Aristotle *Meteorology*	Aristotle *Meteorologica*.
Complete Works	*The Complete Works of Aristotle*, ed. J. Barnes, 2 vols, Princeton: Princeton University Press, 1984, Bollingen Series 71.2.
DK	Diels, H. and Kranz, W. (eds) *Fragmente der Vorsokratiker*, 6th edn, 3 vols, Berlin: Weidmann, 1952.
FHS&G	*Theophrastus of Eresus: Sources for his Life, Writings, Thought and Influence*, eds and trans. W.W. Fortenbaugh, P.M. Huby, R.W. Sharples (Greek and Latin) and D. Gutas (Arabic), together with five others, 2 vols, Leiden: Brill, 1992.
KRS	Kirk, G.S., Raven, J.E. and Schofield, M. *The Presocratic Philosophers*, 2nd edn, Cambridge: Cambridge University Press, 1983.
LCL	*Loeb Classical Library* (London: Heinemann and Cambridge, Mass.: Harvard University Press).
LSJ	Liddell, H.G., Scott, R. and Jones, H.S. *A Greek–English Lexicon*, 9th edn, Oxford: Clarendon Press, 1940, with *Supplement* by E.A. Barber *et al.* 1986, reprinted in 1 vol.
NH	Pliny *Natural History* (*Historia Naturalis*).
NQ	Seneca *Natural Questions* (*Naturales Quaestiones*).
Il.	Homer *Iliad*.
Od.	Homer *Odyssey*.

ACKNOWLEDGEMENTS

It is a very pleasant duty to acknowledge those who have contributed to the completion of this book. It was begun under a grant from the American Council of Learned Societies; I am grateful for the support of this project.

My work has benefited greatly from comments I received when I presented portions of the material covered here at various talks and seminars. Part of what follows is based on a talk given at a conference on the 'Harmony of the Heavens', jointly sponsored by the British Museum and the Warburg Institute, March 1998; a revised version was presented to the Newnham College Research Forum, November 1998, and to the Classics Faculty at the Universiteit Leiden, April 1999. In March 2000 I spoke on 'Ancient Meteorology: Astronomy and Weather Prediction in the Roman Period' at a conference on science and technology at ancient Pompeii at the Deutsches Museum; 'Heavens Above: Tradition and Prediction in Ancient Astrometeorology', for the Horning Endowment for the Humanities at Oregon State University and the Department of the History of Science at University of Oklahoma; and in July 2000 'Stars and Weather Signs: Calendars and Weather Prediction in Ancient Greece and Rome', at the Tanner Symposium on Calendars, Clare Hall. I presented seminars on 'Ancient Meteorology: A Scientific Community Communes with its Past', at the Pudding Seminar, Newnham College (February 2002), and 'Out of the Hands of Zeus: Characterising Ancient Meteorology', at the Centre for the History of Science, Medicine and Technology, University of Manchester (March 2002). I thank these audiences for helpful questions and suggestions. I also thank the Part II students in the Department of History and Philosophy of Science who, during academic years 1998–9 and 1999–2000, read and discussed Epicurus' 'Letter to Pythocles' with me.

I greatly appreciate the help given at various points by the staff of the Cambridge University Library, the Whipple Library, the Classics Faculty Library, the Oriental Studies Faculty Library, Newnham College Library, the Universiteitsbibliotheek, Leiden, and the Antikensammlung, Berlin.

Over a number of years, my colleagues in the Department of History and Philosophy of Science, the Classics Faculty and the Philosophy Faculty at the University of Cambridge, as well as the Department of Classics at the Universiteit Leiden, have been very helpful; I thank them for their advice, encouragement and friendship.

I particularly thank the following individuals, some of whom read and commented on my work, some of whom shared ideas and their own work with me: James Allen, Michael Black, Charles Burnett, Myles Burnyeat, Michael Clarke, Patricia Curd, Andrew Cunningham, Pat Easterling, Resianne Fontaine, Cynthia Freeland, Gerd Graßhoff, John Hall, Jim Hankinson, Robert Hannah, John Heilbron, Richard Hunter, Brad Inwood, Vlad Janković, Daryn Lehoux, Geoffrey Lloyd, Mohan Matthen, Adam Mosley, Alex Mourelatos, Harry Pleket, Marlein van Raalte, Joyce Reynolds, David Runia, David Sedley, Bob Sharples, David Sider, Ineke Sluiter, Heinrich von Staden, Noel Swerdlow, Karin Tybjerg, John Vallance and an anonymous reader. Several others contributed in various ways, including helping to obtain images: Robert Anderson, Lindy Divarci, Clare Drury, Janet Dudley, Marina Frasca-Spada, Tamara Hug, Boris Jardine, Andreas Kleinert, Diana Lipton, Peter Lipton, Anna Mastrogianni, Lisa Newble, Ruth Phillips, Christine Salazar, Kemal de Soysa, Laurence Totelin, Catriona West, Frances Willmoth and Rosemary Yallop. I would like to thank Nicholas Branson for introducing me to John Hall, and Richard Stoneman, Catherine Bousfield, Linda Paulus and Celia Tedd at Routledge for their helpful advice and forbearance, Sarah Moore for her patience and good humour and everyone at Florence Production. Paul Cartledge very generously offered to read page proofs and made a number of welcome comments and corrections.

I am grateful to Roger French for his encouragement and interest in this project; I wish that I had been able to share the finished product with him.

Finally, I note my gratitude to my friends, who probably had to talk more about the weather than they ever intended to; my parents, who are always very supportive and encouraging; and my husband, Niall Caldwell, who contributed to this work in ways too numerous to list (here).

1

ANCIENT METEOROLOGY IN GREECE AND ROME: AN INTRODUCTION

Today, for many of us, the word 'meteorology' conjures up images of the evening news and the weather forecasts offered by television 'meteorologists'. Given the current debates regarding global warming and climate change, the term 'meteorology' points to a science concerned with controversial and difficult-to-prove hypotheses. For many people, meteorology appears to be a speculative subject, in which predictions are very often not borne out. Indeed, current chaos theorists emphasize the difficulty (if not impossibility) of making accurate weather predictions. Nevertheless, or perhaps because of a shared sense of frustration, the weather is always a 'safe' subject for conversation, for we assume that everyone is interested in it. Indeed, a certain preoccupation with it is regarded in some regions and occupations as completely normal.

We often have the impression that meteorology is primarily concerned with weather prediction, yet it is not only concerned with forecasting, but also with the explanation of weather phenomena. (Indeed, some would argue that it is difficult or impossible to predict without understanding the causes of weather.) Today, understanding the causes of weather and climate is regarded as necessary, in order to cope with phenomena like El Niño and to prevent global warming. And while the relationship between the ability to explain and to predict natural phenomena is not always straightforward, we tend to regard the ability to predict as a hallmark of modern science.

This book focuses on ancient Greek and Roman approaches to the prediction of weather and the explanation of the causes of meteorological phenomena. In ancient Greece and Rome, the study of meteorology covered a broader range of phenomena than is usual today, and was of interest to many, including farmers, poets, philosophers and physicians. The modern term 'meteorology' comes from the ancient Greek word *meteōrologia*, which refers to the study of

1

the *'meteōra'*. But the answer to the question 'what does the term *meteōra* cover for the ancients?' is not straightforward, for not all the authors considered here agree. Some, for example Plato, used related terms to refer to thinking about 'lofty' things, but it is not entirely clear what these 'lofty' things were.[1] We might expect that ancient meteorology would focus on things high in the atmosphere. When we look at the ancient Greek and Roman texts on meteorology, we discover discussions about 'lofty' things which today would be regarded as astronomical phenomena, such as comets and the Milky Way. However, earthquakes and other phenomena that would today be regarded as geological and seismological were also treated in ancient texts on meteorology. And, as we would expect from our modern term, much of ancient meteorology too was concerned with weather.

Given the predominance of farming in ancient society, it is not surprising that Greek and Roman authors wrote a great deal about the weather. Furthermore, the movement and transport of agricultural stock and manufactured products were vital in the ancient world. Those who had to organize such movements were, of necessity, much concerned with weather. But the study of ancient meteorology was not simply a practical matter. Many ancient authors who wrote on the subject were aiming to address important questions about the nature of the world and how we understand it, questions about whether or not the cosmos is a unity, and what sort of explanations of the cosmos (and its parts) are possible.

While the main aim of this study is to provide an overview of ancient Greek and Roman approaches to the prediction and explanation of meteorological phenomena, there are several 'sub-plots' underlying the work. One is a concern with the forms in which these approaches have been set down, described and disseminated. Because of the diverse character of the ancient texts, I devote some attention to questions of format and genre. Sometimes, because of our own implicit expectations regarding modern scientific communication, we are surprised by the range and diversity of the forms of communication of scientific ideas and methods in other periods and cultures.

Poetry plays an important role as a central genre for sharing meteorological information, and not only in antiquity. The use of poetry as a way of communicating scientific ideas may be surprising to modern sensibilities. Several major Greek and Latin poets (including Hesiod, Aratus and Lucretius) composed poems that included, in some instances, detailed meteorological material; some of these poems (including that of Aratus) were verse treatments of prose treatises. And prose authors of learned and technical treatises incorporated quotations from and allusions to these poems. In other words, authors of didactic prose treatises considered the poets to be authoritative sources of information on meteorological matters.

Questions relating to format and genre can also be revealing with respect to the relationship of authors to their predecessors. In many cases our primary sources are fragmentary, incomplete, or lost, and our understanding of ancient meteorology is derived from ancient authors reporting the ideas, methods, or observations of others. This book is based on an attempt to reconstruct, through ancient accounts, a brief (and not complete) history of ancient meteorology. Of course, those ancient authors who preserved and reported the work of others did so for their own purposes, which did not include the production, in the early twenty-first century, of a history of ancient meteorology. Recognizing this, while I have tried to query the purpose behind each author's approach and 'motives', I realize the limitations of trying to discover and understand any author's 'agenda'.

By paying attention to the issues raised by the genre and format of ancient works on meteorology and also to questions regarding the value, usefulness and reliability of information contained in the works of others, I intend to signal that these ancient works display interesting tensions regarding the status of authorities and the use of knowledge derived from them. These tensions are deeply embedded in the cultures and values of the Greco-Roman world and contribute to the rich complexity of ancient projects to predict and explain meteorological phenomena.

Insights into ancient notions about the weather are found in the earliest extant ancient Greek works, the Homeric and Hesiodic poems, which include many references to meteorological phenomena. Of course, the primary purpose of the ancient poets was not to explain and predict weather phenomena, but the poems do give both incidental and deliberate indications of archaic Greek ideas about the weather, and the use of such knowledge. The audiences of the poems were probably not particularly interested in meteorology, but the descriptions of weather phenomena would have carried familiar meaning. Furthermore, the poems of Homer and Hesiod serve as points of reference in later discussions of meteorology. Both poets are echoed and quoted by numerous later authors on meteorology. Hesiod, in particular, is mentioned in relation to prediction; references to Homer sometimes feature in explanations of weather.[2]

The authorship and dating of the epic poems have been debated since antiquity (beginning at least in Alexandria), but there is now

Figure 1.1 Bronze statuette of Zeus with thunderbolt, *c.* 470 BCE, found at
Dodona (ancient oracular temple and sanctuary of Zeus).

Source: Staatliche Museen zu Berlin-Preußischer Kulturbesitz Antikensammlung.

general consensus that the *Iliad* and the *Odyssey*, both attributed to
Homer, were composed in the second half of the eighth century BCE.
The *Iliad* is thought to be older than the *Odyssey*; the first is dated to
about 750 BCE and the second to about 725 BCE. Little is known
about the poet known as Homer and it is a question whether the
same author composed both poems. Interestingly, the description of
weather phenomena has led some to locate the author of the *Iliad* on
the eastern side of the Aegean; the poet's description of the winds
suggests familiarity with regions east of Thrace.[3] The *Theogony* and
the *Works and Days* are both generally ascribed to Hesiod. From the
fifth century BCE onwards, scholars disputed which of the two epic

poets, Homer or Hesiod, was the older. A date for Hesiod of around 700 BCE is now accepted.

In the Homeric and Hesiodic poems, meteorological phenomena are often linked to the activities of gods. Many meteorological phenomena are personifications or epiphanies of gods, or are sent directly by Zeus. In the *Iliad* and the *Odyssey*, Zeus, sometimes referred to as 'cloud-gatherer' (*Od.* 5.21; *Il.* 1.511), is often responsible for meteorological events.[4] Rain comes from Zeus; at *Il.* 12.25 Zeus rains continually. He produces thunder (*Od.* 14.305) and hurls bolts of lightning (*Il.* 8.134; *Od.* 12.415). Zeus is described as carrying lightning in his hand (*Il.* 13.242); he causes storms (*Od.* 9.67; see also *Il.* 15.379; *Il.* 16.385f.); and places rainbows in the clouds as a portent for humans (*Il.* 11.27–8). But other divinities can also produce meteorological phenomena; for example, together, Hera and Athene cause thunder (*Il.* 11.45). Winds may come indirectly from Zeus (*Il.* 2: 144–6, where Euros and Notos raise the waves, springing from the clouds of father Zeus); winds can also be controlled by various gods. In the *Odyssey* (*Od.* 5.382–5) Athene is able to control the wind:

> But now Athene, daughter of Zeus, planned what was to
>   follow.
> She fastened down the courses of all the rest of the
>   stormwinds,
> and told them all to go to sleep now and to give over,
> but stirred a hastening North Wind, and broke down the
>   seas . . .[5]

The goddess Kalypso sent a wind to carry Odysseus across the sea (*Od.* 5.167). Poseidon, often described as 'earth-shaker' (*Od.* 1.74, 5.423), also pulled together clouds and let loose a storm of winds (*Od.* 5.291–6). As the 'earth-shaker', Poseidon has control over not only seismological but also meteorological phenomena. At one point in the *Odyssey* (*Od.* 10.19ff.), we learn that Zeus put Aiolos, a mortal, in charge of a bag containing all the winds. Aiolos stowed the bag carefully, tied up with a silver string, letting out only the West Wind to blow ships safely on their course. But greedy men were curious about the contents of the bag. Believing it might contain silver and gold, they opened it, releasing all the winds, resulting in a storm which swept them away. Control of the weather has great symbolic value in the Homeric poems.

The Hesiodic *Theogony* presents a lengthy and complicated genealogy of the gods, in which a mythological account of meteorological phenomena is crucially embedded. Zeus, son of Kronos and Rhea, is the storm-king, the cloud-gatherer. Following various exploits, with the help of the Hundred-Handers he drives the Titans from heaven:

> And now Zeus no longer held back his strength.
> His lungs seethed with anger and he revealed
> All his power. He charged from the sky, hurtling
> Down from Olympos in a flurry of lightning,
> Hurling thunderbolts one after another, right on target,
> From his massive hand, a whirlwind of holy flame.[6]

Following this, Gaia (Earth), pregnant by Tartarus, gives birth to Typhoeus, and we then get a good deal of weather and winds; Typhoeus is the source of winds which cause shipwrecks and destruction.[7] The Homeric and Hesiodic gods play crucial roles in causing and controlling meteorological phenomena.

Some would discount mythological accounts as being of no interest from a 'scientific' standpoint. But the whole question of what is 'scientific' is not agreed upon by historians and philosophers of science, least of all for the ancient period. Some argue that the only ancient texts worthy of the appellation 'scientific' are mathematical works, yet most historians of science, as well as historians of philosophy, would agree that the ancient writings on natural philosophy are properly studied as part of the history of science. Whether or not that makes those writings 'scientific' is another matter. The relationship between traditional mythology and ancient philosophy is complicated, as the ancients themselves acknowledged.

Myth was regarded by many ancient philosophers, including Plato, as an acceptable form of description and explanation. Both Plato (*Cratylus* 402b; *Theaetetus* 152e, 180c–d) and Aristotle (*Metaphysics* 983b27f.) suggested, perhaps jokingly, that Homer and Hesiod were the fathers of ancient philosophy. W.K.C. Guthrie noted that 'Plato was fond of calling Homer the ancestor of certain philosophical theories because he spoke of Oceanus and Tethys, gods of water, as parents of the gods and of all creatures'.[8] While such pronouncements may have been made lightly, there may also have been an element of seriousness. Within some definitions of mythology, the accounts of the traditional gods and their meteorological activities may be understood, quite reasonably, as a form of explanation. There was an ancient tradition of regarding the earliest poets as intellectuals; standing at the fountainheads of tradition, they helped shape intellectual agendas. While ancient philosophers were by no means unanimous in their views of the early poets, the variety of reactions to them indicates a perceived need to engage with their accounts.[9]

Certainly, many meteorological phenomena, including storms, lightning and thunder, are potentially dangerous and frightening. Many practical activities (including farming and seafaring) are, to some extent, determined and circumscribed by meteorological events. The use of mythology to explain this group of phenomena and the reliance on signs and omens to predict their occurrence are an important part of the fabric of many cultures, including several in antiquity. The desire to account for and to predict weather phenomena is shown in the literature that survives. In addition to early mythological accounts and omen literature, there is a large body of literature that spans the ancient period and is concerned with explanation and prediction of meteorological events. This literature, produced by some of the most influential writers of antiquity, falls into two distinct traditions: one philosophical and largely explanatory, the other for the most part predictive and linked to observation (and recording) of phenomena, including astronomical events and animal behaviour.

While it is always difficult to determine the origin of an idea or methodology, the importance of the Hesiodic poems and their influence on later ancient authors writing about meteorology were clearly significant. The *Works and Days* stands at the beginning of a special tradition of astrometeorological texts, which correlate astronomical phenomena with the seasons and with weather events. To a large extent, the final section of the *Works and Days* may be regarded as an early farmers' almanac.[10] In the Homeric poems, there is a sense of an annual cycle of recurrent seasons ('the year had gone full circle and come back with the seasons returning', *Od.* 10.467f., 11.294–5; *Il.* 2.550). Several seasons are mentioned: winter, spring, summer, late summer/early autumn.[11] These seasons are marked in various ways: in some, particular sorts of weather are characteristic. The *Iliad* 16.384–5 describes the late summer/early autumn (ὀπώρη) as a time of violent rain and flooding. Winter (χειμών) is a time of unceasing rain (*Il.* 3.4) and stormy weather (*Od.* 4. 566, 14.522). Spring (ἔαρ, *Il.* 6.148) is windy, and also the time of the nightingale's song (*Od.* 19.519). The Homeric poems give some slight evidence of a relation between astronomical phenomena and agricultural seasons: late summer/early autumn (ὀπώρη) is referred to as the time of the star known as Dog-of-Orion (*Il.* 22.29), which is also a time of harvest (*Od.* 11.192) and fever (*Il.* 22.30–1). However, the link between an astronomically based calendar and agricultural activity is not well developed in the Homeric poems and certainly not to the extent that is evident in Hesiod's *Works and Days*.[12]

The ancient practice of using the risings and settings of bright stars to mark out the seasons and to predict weather phenomena was based on ideas that the motions of the heavens are related to terrestrial events, particularly weather.[13] Astrometeorological *parapēgmata* (lists of star phases and associated weather predictions) and the astrological literature, including Ptolemy's (second century CE) *Tetrabiblos*, are examples of the astronomical tradition of weather prediction, which will be examined in the next chapter. This tradition survived through late antiquity and was developed further in the medieval period, in both the Arabic and Latin traditions, and continued well into the early modern period.

Weather prediction was useful not only for farmers, but also for physicians. For example, the author of the Hippocratic medical treatise on *Airs, Waters, Places* states at the outset that physicians must be familiar with the differences between the seasons, so as to know what changes to expect in the weather and to predict the effects on the health of themselves and their patients.[14]

In addition to the ancient mythological accounts and the traditions of weather prediction, there are other writings which attest to interest in talking about the weather. Herodotus devoted a lengthy section of his *Histories* (part of Book 2) to a description of the Egyptian weather and climate, as well as other meteorological topics. Aristophanes, in the play *The Clouds* (331ff.), humorously criticized the new learning of Socrates and other 'sophists' who, he suggested, were 'always talking about clouds and things'.

Natural philosophers, from the Presocratics through the Hellenistic period, discussed the causes and offered explanations of weather phenomena. Sadly, only fragments (in some cases lengthy) and the titles of Presocratic philosophical texts survive. Fortunately, Aristotle's lengthy treatise, the *Meteorology*, provides information not only on his own ideas, but on those of his predecessors as well. Of later writers, Lucretius' *On the Nature of Things* and Seneca's *Natural Questions* provide information on the Hellenistic philosophical schools, and other writers, including Pliny the Elder, treat the subject as well. These authors will be discussed in detail in the final three chapters.

It is important to note that many ancient authors wrote about meteorology within the framework of a work conceived for a broader purpose; the explanation of meteorological phenomena was not always the principal aim. For example, in the first century BCE,

Lucretius gave an account of the physical theory of his predecessor Epicurus (341–270 BCE) in the Latin poem *On the Nature of Things* (*De rerum natura*). Lucretius sought to eliminate fear of the intervention of gods in the natural world; as part of this project, he explained various meteorological phenomena. Similarly, it was within a larger programme designed to aid moral improvement that the first-century CE Roman author Seneca, in the *Natural Questions,* treated meteorological topics. Notably, works by these authors were accessible and influential in the medieval and early modern periods.[15]

Given the considerable interest in meteorology in antiquity, it is surprising that this subject has received so little attention from modern scholars. Otto Gilbert's *Die meteorologischen Theorien des griechischen Altertums*, published in 1907, remains the most recent attempt at a comprehensive treatment. Gilbert treated 'element' theory first, moving from the evidence of folklore through to the Stoic philosophers; he then turned to consider particular meteorological topics. His interest in the theory of elements pervades the work. Charles Kahn, in *Anaximander and the Origins of Greek Cosmology* (1960), criticized Gilbert's approach, arguing that 'by projecting the philosophic theory of the elements back into Homer and the "Volksanschauung" [popular world-view], he rendered the historical sequence of ideas unintelligible'.[16]

Kahn judged ancient meteorology to be a rather conservative enterprise, lacking in significant development once certain central ideas had been put forward. In his view, the idea of a 'vital link between the study of the heavens and that of living things upon the earth' was at the heart of the view of nature suggested by the Milesian philosophers, particularly Anaximander. He suggested that 'it might be possible to show that all the meteorological theories of antiquity represent only minor variations' on this theme. In arguing for the conservative nature of Greek and Roman meteorology, Kahn suggested that:

> Once such a rational theory had been expounded, there was little need for a new one. The goal of a naturalistic explanation had been reached; the most impressive of atmospheric phenomena no longer revealed the hand of an anthropomorphic god. The attention of later thinkers was thus diverted to new problems, or restricted to a rearrangement of the details. Hence the conservative character of Greek meteorology, and hence the permanent sway of Milesian doctrine throughout antiquity.[17]

9

I would argue that, while in many ways Greek and Roman explanations of meteorological phenomena tended to share certain themes, such as the assumption of a liability to rational understanding and a link between the heavens and the earth, the variations developed on these and other themes are not insignificant or merely restricted to rearranging details. For example, several ancient authors display a keen interest in discussing the methodology they have employed in their meteorological works, and deliberately confront the problems of achieving adequate empirical access to a number of meteorological phenomena.

Yet, to some extent, the study of meteorology, as practised by ancient Greeks and Romans whose texts survive, does appear to have been a conservative tradition, in which tradition itself and conservatism were both valued. But meteorology was by no means a monolithic study; rather, meteorology actually incorporated several, sometimes overlapping, 'traditions', including poetic as well as philosophical practices and conventions. Nor were all explanations and predictions merely recycled from earlier writers. Rather, the deliberate engagement with earlier thinkers and writers is a signal characteristic of ancient meteorology.

The earliest surviving Greek texts, the Homeric and Hesiodic poems, show that meteorological phenomena were traditionally linked to the gods, often as their epiphanies. Throughout the history of ancient Greek and Roman meteorology, the authority of these earliest poets was valued. Even though in some authors the links between the gods and weather phenomena were severed, in favour of other, naturalistic, explanations, the poets were still revered as offering relevant knowledge. Some have argued that Greek natural philosophy excluded the gods as agents in the natural world; however, this so-called 'exclusion' is an oversimplification. Certainly, in most of the ancient philosophical schools there was an assumption of some 'divine' presence in the cosmos; in some cases, the cosmos itself was thought to be divine. The nature of this 'divinity' (or 'divinities') was a central concern for many ancient philosophers, but it is important to recognize that what is 'divine' was not necessarily understood as being a 'god'. And, in those philosophical schools in which the existence of gods was accepted, the role of these gods was subject to debate. Questions about whether the gods play a role in causing meteorological phenomena recur in both Greek and Latin texts.[18] There is a strong sense of the desire of those interested more generally in cosmology and theology to continue to confront issues relating to the fears and sense of disorderliness associated with the traditional gods and their control of the *meteōra*. Those texts which deal with prediction and explanation of meteorological phenomena may be understood as attempts to gain some measure of control over these important, and often frightening, events.

Throughout antiquity, poetry was an important medium for the communication of many types of information, including technical data and instructions. While prose forms often predominated in the technical literature, the number of widely read poems communicating meteorological information is striking. The long-lived authority of Homer and Hesiod on meteorological issues is attested by the many quotations and allusions to their poems, even in prose treatises. The extent to which the poets serve as sources of knowledge for technical writers is worth noting; on the topic of stars and weather signs, certain writers, including Columella, Pliny and Seneca, considered the poets Hesiod and Virgil to be as valid as some specialist astronomers.

Hesiod may be regarded as standing at the beginning of a long-lived approach to predicting weather, which depended on correlating weather conditions with celestial events. While other techniques were also employed, weather prediction was routinely presented as a practice whose methods had been pursued over a long period of time, crossing geographical, cultural and linguistic boundaries in the ancient Mediterranean world. Techniques for predicting weather were part of a scientific tradition in which tradition itself was valued. Modifications are not presented as innovations, but as refinements.

Approaches to the explanation of meteorological phenomena mostly follow well-trod paths. The difficulty of dealing with meteorological phenomena, whether predicting or explaining them, is a recurrent theme. Aristotle goes so far as to warn, at the beginning of his treatise, that meteorology is a very difficult subject: some phenomena are inexplicable, others can only be partially understood. The difficulty of gaining proper access to distant phenomena is cited, and many authors employ analogies with easily accessible everyday experience as a means of understanding and explaining the causes of certain phenomena. Quoting and commenting on the opinions of reputable and well-regarded predecessors is another approach frequently employed: while these opinions were by no means always accepted, their inclusion in the discussion is characteristic of many accounts. Many authors were content to offer different possible explanations for the same meteorological phenomenon, rejecting the idea that only one explanation should be offered; here, the opinions of others could be used as alternative explanations. In some cases, this was because meteorological phenomena were regarded as in principle difficult to access; therefore it was regarded as unrealistic to limit the possible causes offered as explanations.

The reliance on the work of others, in shaping both predictions and explanations, is a key characteristic of Greek and Roman meteorology. The authors of the ancient texts deliberately incorporated the information, the ideas and opinions from a number of sources. The authors were writing for an audience, which sometimes included friends, colleagues and students. In some cases, the intended audience may have been members of the same philosophical school; others suggest that the audience was meant to be 'converted' to the author's way of understanding. Coping with meteorological events was a sort of shared 'project', motivated, no doubt, by a number of concerns and desires, especially understanding frightening and dangerous phenomena. But it is not at all clear to what extent the meteorological works as we have them would have been encountered by potential users such as farmers and sailors; nor is it clear how much help they would have been. The experience and observations of such people were incorporated into the ancient texts, to varying degrees. This is not particularly surprising, given the tendency of members of particular philosophical schools to reflect on and develop the ideas of the founders; similarly, within astronomy and astrometeorology, observations made centuries earlier were still useful. Meteorology is an area in which Greek and Roman authors worked within shared traditions, forming a community of contributors and collaborators that spanned many centuries. The authors writing on meteorology exhibit a close engagement with the past, demonstrated by the incorporation and utilization of records, observations, ideas and approaches contributed by long-dead predecessors. Those providing meteorological predictions and explanations are part of an imagined community, which relies (in a variety of ways) on the continuing engagement of others interested in the same problems, even though those other participants may no longer be alive. Because of this, the contributions of those who lived before continued to be valued, even when not fully accepted. Indeed, ancient meteorological approaches were surprisingly long-lived even beyond antiquity, and survived in some forms in much later cultures. 'Farmers' almanacs' arguably owe their existence to ancient and traditional models.

In what follows, the authors discussed cover a wide span of time, from the eighth century BCE to the sixth century CE; furthermore, they lived in different places. It would be misleading to suggest that all of the authors here shared the same understanding of the physical

nature of the world. Nevertheless, there are some ideas which underlie much of ancient Greek and Roman cosmology, which are worth
summarizing here. When the word 'cosmology' is defined as the
study of the structure of the universe, 'cosmos' refers to the universe
as a whole. In a more general sense, the Greek word 'cosmos' (κόσμος)
means 'order' or 'adornment', and is related to the English word
'cosmetic'. This sense of order and aesthetic beauty underlies the
Greek view of the universe.[19] Several of the authors considered here,
for example Aristotle, discussed cosmology in detail; others did not
articulate their views. Cosmic order is described in a number of ways;
in certain authors, the idea of the harmony (or fitting together) of the
universe is crucial. For some, the earth is regarded as a living body,
or as like one. That the universe is spherical, that the earth is in the
centre of the universe, and that the heavens move spherically around
the earth were ideas which were widely held among those interested
in such questions. That there are fundamental 'elements' which
underlie the composition of the material substances in the universe
and that these substances are subject to certain types of change
(including motion) were ideas which were shared by many, even
when they debated the nature of the 'elements' and the types of
change they undergo.[20]

About the phenomena: The climate of the Mediterranean region
is fairly even, characterized by mild, rainy winters and dry, hot summers. The geography of the ancient Greek and Roman world covers
a multitude of different types of landscape. In some cases, a particular landscape may be of very limited extent. There are important
contrasts between low-lying and mountainous regions, and many
micro-climates. In the lowlands, rainfall occurs unevenly. Torrential
downpours are frequent in winter, while summers tend to be very dry
but there is great variability in interannual rainfall. In mountain
regions an alpine climate prevails, with harsh winters and warm summers. While climate change during historical periods has been
recorded in some regions, it has never been demonstrated that the climate of the ancient Mediterranean world was fundamentally different
from that of today. It seems reasonable to assume that the ancient
writers were describing phenomena which would be familiar to those
currently in the same geographical region.[21]

2

PREDICTION AND THE ROLE OF TRADITION: ALMANACS AND SIGNS, *PARAPĒGMATA* AND POEMS

In ancient Greece and Rome, a number of methods were used to predict meteorological phenomena. The variety of works conveying information about meteorological prediction is striking, and includes writings in several literary forms. There are prose treatises which discuss meteorological prediction, but two other genres are particularly important here: astrometeorological calendrical texts and didactic poems.[1] Some of our sources are entirely devoted to the topic, for example the work known as *On Weather Signs*, discussed below. In other cases, works which focus on different subjects nevertheless incorporate information about predictive techniques which were used; this is the case with many of the didactic poems, which list various sorts of 'signs' useful for predicting weather. As we shall see, many of the authors of the texts that deal with weather prediction expressly make reference to earlier writers. Furthermore, there are certain texts which are clearly central to Greek and Roman ideas on weather prediction. In particular, the Hesiodic poem *Works and Days* (especially the 'calendar' or 'almanac' at the end of the work) and the treatise *On Weather Signs* appear to provide a foundation for later writers interested in meteorological prediction. As David Sider has pointed out, the relevant section in the *Works and Days* focuses primarily on regularly occurring seasonal weather, which can be set out on an annual basis; *On Weather Signs* is chiefly concerned with indications of weather changes which appear relatively shortly before the weather events happen, even if those events typically occur in a particular season.[2] The *Works and Days* stands at the head of a long tradition of texts concerned with weather prediction, but it must be acknowledged that prognostication was not a key aim of the work.

Astrometeorology

While it would be wrong to give the impression that all ancient weather predictions were based on correlating astronomical phenomena with weather events, nevertheless, such correlations are part of an important and long-lived technique, practised in several cultures. For example, Babylonian scribes of the second millennium BCE were responsible for the great omen series known as *Enūma Anu Enlil*, representing a vast collection of celestial omens, regarded as signs from the gods. Such omens were important in political, military and agricultural affairs, and were consulted by advisers to the Assyrian kings. About 7,000 omens were collected in seventy cuneiform tablets; these include lunar, solar and meteorological, as well as stellar (and planetary) omens. The omens took the following form: 'If Jupiter stands in Pisces: the Tigris and the Euphrates will be filled with silt'; 'If the moon is surrounded by a halo and the Bow star stands in it: men will rage, and robberies will become numerous in the land.'[3]

From the eighth or seventh century BCE until the first century, scribes were engaged in nightly observations of the sky; they recorded the dates and locations of the ominous phenomena of the moon and planets with reference to the stars and constellations. Noel Swerdlow, in *The Babylonian Theory of the Planets*, explained that 'the systematic collection of celestial and meteorological omens as carried out in Mesopotamia required a systematic study of celestial and meteorological phenomena'. The observer-scribes discovered that the risings and settings of the planets are recurrent phenomena that follow cycles in which the same phenomena occur on nearly the same dates of the calendar month. The scribes of *Enūma Anu Enlil* apparently began their systematic observations of celestial phenomena during the reign of Nabonassar (747 to 733 BCE), recording both astronomical and meteorological phenomena in reports known as 'regular watching' (*nasaru sa gine*). These reports are known as Astronomical Diaries; the earliest surviving diary is from 652 BCE. It is not known how long before that date the observations began to be taken and recorded. The Diaries are records of astronomical observations, but they also report meteorological phenomena. Swerdlow has suggested that, in fact, 'celestial and meteorological [phenomena] were probably considered to be related or even one and the same'. He notes that the most extensive and detailed reports contained in the Diaries are actually of the weather.[4]

Of course, weather determines the visibility of astronomical phenomena; while records of weather phenomena constitute a continuous meteorological record extending over at least 600 years, this may not have been the primary purpose of the scribes. In fact, only bad weather is reported; if it is clear and warm and dry and calm, there is no weather report. Swerdlow explains that 'it appears from the Diaries . . . that the weather in ancient Mesopotamia was frequently terrible, frustrating the efforts of the most devoted watcher of the heavens, with night after night of clouds and rain of various sorts, described in detail by numerous technical terms, as well as fog, mist, hail, thunder, lightning, winds from all directions, often cold, and frequent "*pisan dib*", of unknown meaning but always associated with rain'. He notes that such empirically derived weather records served scientific purposes, allowing weather events to be related to other celestial phenomena, with the goal of making forecasts. Furthermore, because the weather itself is ominous, specific meteorological events (including winds, rainbows and cloud cover) were regarded as significant and useful for prediction and divination.[5]

Celestial omens were systematically collected since the beginning of the first millennium BCE. Such omens often had the form that the protasis (the 'if' clause of a conditional statement) collected observations of both astronomical events and meteorological events. The apodosis (the 'then' clause) referred to political or other events concerning the fate of the state or its rulers. The relationship between astronomy and meteorology was not of cause and effect; rather, both astronomical and meteorological events were treated as the same sort of triggers for omen. Gerd Graßhoff has argued that the Babylonian Astronomical Diaries were, in fact, the observation-records of the astronomers who also developed the theoretical models. The Astronomical Diaries show the empirical procedures used to gather knowledge of both astronomical and meteorological phenomena. Graßhoff believes that correlations of stellar phases and weather phenomena were derived either from the older omen literature or from the Diaries themselves.[6]

The Babylonian approach to dealing with celestial and meteorological phenomena became known to Greeks by at least the third century BCE.[7] There was, however, within ancient Greece and Rome, an apparently independent tradition, in which the practice of astrometeorology appears to have depended on a relatively well-articulated notion of cosmic harmony, in which the celestial influences the terrestrial.[8] The linking together of the celestial and terrestrial into a harmonious universe was crucial to the tradition of predicting some meteorological events; in effect, ancient Greek astrometeorology depended quite literally upon a harmony of the cosmos.

The Greek word *harmonia*, in its most basic sense, refers to a 'means of joining or fastening'. Thus, in the *Odyssey* (5.248, 5.361) a form of the word is used to refer to the joining together of the timbers of Odysseus' raft. Other uses, perhaps extensions of the original sense, refer to a framework like the human form, to a covenant or agreement, to order and structure, and to a method of stringing musical instruments to fit a musical mode or scale. The use of *harmonia* specifically referring to the heavens or universe seems to date from the fifth century BCE, and to have been Pythagorean in origin.[9] The earliest surviving use of the term to refer to cosmic harmony may be found at the end of Plato's (*c.* 429–347 BCE) dialogue *Timaeus*, where the Pythagorean Timaeus explains that: 'the motions which are naturally akin to the divine principle within us are the thoughts and revolutions of the universe'.[10]

That the whole universe is joined together by cosmic motions is an idea which underlies the whole work. Because the cosmos is a whole, whose parts fit together in a harmonious way, we (as beings created as part of that universe) can attain the best life possible by gaining knowledge of these harmonies and revolutions. In the *Timaeus*, *harmonia* appears to be used in at least two senses, which are crucially related. *Harmonia* refers to the cosmic fitting together and to the divine sound of music that can be heard by mortals, allowing them to imitate the cosmic harmony and to be at one with the universe.

Timaeus had higher ends in mind, but it should be noted that the idea of the benefit and usefulness of the study of the motions of the heavens and of the fitting together of the universe was not new to either Plato or the speaker in the *Timaeus*. For example, the first century CE Roman author Pliny the Elder (*NH* 18.58.273) credits Democritus (fl. 430 BCE) as the first to realize and point to the relationship between the heavens and the earth. And in the fourth century BCE Aristotle (*Politics* 1259a6ff.) related a story of how Thales (fl. 586 BCE) turned a sizeable profit by monopolizing oil presses, based on his study of the heavens.

It is in one of the earliest Greek texts to come down to us, the Hesiodic poem *Works and Days*, that we find the first suggestion that knowledge of the motions of the celestial bodies may be useful, both for practical purposes and for ethical ends as well. Throughout the *Works and Days* it is made clear that practical measures should be employed to overcome the evil of the world.

The poem is composed of four parts.[11] The first describes the origin and subsequent spread of evil, while the second explains how people may escape these evils through work, especially in agriculture and trade. A series of maxims useful in everyday life comprises the third section. The final section lists the days of the month which are favourable for industry and agriculture.[12] The four parts of the poem are linked by their single aim: to show how to live in a difficult world. Evil may be overcome through hard work; familiarity with astronomical events allows one to know which times in the year will be particularly appropriate and favourable for various tasks. The poet describes the end of winter and beginning of spring at lines 564ff.:

> Now, when Zeus has brought to completion
>   sixty more winter
> days, after the sun has turned in his course,
>   the star
> Arcturus, leaving behind the sacred stream
>   of the ocean,
> first begins to rise and shine at the edges
>   of evening.
> After him, the treble-crying swallow,
>   Pandion's daughter,
> comes into the sight of men when spring's just
>   at the beginning.

And, at 597ff., he gives the following instruction to:

> Rouse up your slaves to winnow the sacred yield
>   of Demeter
> at the time when powerful Orion first shows himself;
>   do it
> in a place where there is a good strong wind,
>   on a floor that's rounded.

He urges (609ff.):

> Then, when Orion and Seirios are come to the middle
> of the sky, and the rosy-fingered Dawn
>   confronts Arcturus,
> then, Perses, cut off all your grapes, and bring
>   them home with you.[13]

The poet is not merely offering the general suggestion to use the stars to tell the season. Rather, the level of detail here makes it clear that, probably over a lengthy period of time and through shared effort, ancient farmers had determined a calendar to guide their agricultural activities through the course of the year. Here the poet is sharing the fruits of this labour.

Astronomical almanacs and the weather

The *Works and Days* is partly a farmers' almanac, in verse form, with instructions on when to do what, with some technical advice thrown in. Clearly, by being familiar with astronomical events, farmers would know which times in the year would be particularly appropriate and favourable for various tasks. In some cases, Hesiod indicates that celestial phenomena can be associated with specific seasonal weather. So, for example, 'when the Pleiades plunge into the misty sea to escape Orion's rude strength, then truly gales of all kinds rage' (619ff.; at the end of October or the beginning of November); and 'fifty days after the solstice . . . the season of wearisome heat is come to an end' (663f.).[14] In the *Works and Days*, the almanac is, for the most part, astronomically based; an exception is at 448, where the poet suggests that the annual cry of the crane signals the beginning of planting and winter rains.[15] Hesiod's *Works and Days* stands at the beginning of a special tradition of Greek and Roman astrometeorological texts,[16] which correlate astronomical phenomena with weather phenomena. But it should be recognized that the poetic farmers' almanac is itself situated within a wider-ranging ethical work, which explains how to behave and what to do to improve life.

In the early part of the last century, stone fragments of two *parapēgmata* texts were discovered at Miletus, one dated to the late second, the other to the early first century BCE.[17] Prior to their discovery, only a literary form of *parapēgma* was known, and the dating of one of the most important literary examples (the so-called 'Geminus *parapēgma*') is far from certain.[18] It seems that originally the term *parapēgma*, which means 'something on which you fix something next to something else',[19] described an inscribed stone which was displayed for public use, rather as stone sundials were sometimes displayed in public places.[20] Archaeologists have discovered a small number of inscribed stones, mainly fragments, which have been described as *parapēgmata*.[21] The *parapēgma* had holes beside the text, in which a peg could be inserted. There are scattered literary references to *parapēgmata* and surviving stone fragments which have been called *parapēgmata*; in some cases it is not entirely clear what the peg is marking or counting.[22] Some of the surviving *parapēgmata* are clearly astronomical; an even smaller number of fragments seem to relate astronomical events to weather conditions. The Milesian *parapēgma* fragments have peg-holes provided in order to represent each day. A peg was placed in the hole for the appropriate day, so that one would not lose the place in the month. The text associated with each peg-hole (or day) provided relevant astronomical (and, in some cases) meteorological information.[23]

There were many different calendars in use in the Greek and Roman world; for example, each Greek city-state had its own calendar. Early Greek calendars, like most other calendars of Mediterranean peoples, were lunar. Furthermore, the calendars of each *polis* tended to be somewhat fluid; if necessary, adjustments were made to ensure that various political or religious events would occur on the desired day.[24] The civil calendars of the Greek city-states tended to be somewhat chaotic (and were often the butt of jokes),[25] but the *parapēgma* offered regularity based on celestial phenomena. The astronomical *parapēgma* may have been the invention of Meton and Euctemon, in the second half of the fifth century BCE in Athens.[26] The construction and provision of a stone *parapēgma* would have been a laborious and costly procedure; when considering the stone *parapēgmata* it is worth remembering the investment they represented. Those fragments that survive indicate the relatively large size of the *parapēgmata*; this emphasizes the very public character of the inscriptions.[27]

At this point it might be helpful to provide an example of these texts; the following is from the earlier Milesian inscription (456B, middle column, according to the text of Diels and Rehm). (Each ○ indicates where a drilled hole appears on the stone inscription, each presumably indicating one day to be counted in sequence; the horizontally arranged holes have no corresponding text applying to that day.)

○ The Sun in the Water-Pourer [Aquarius].
○ <The Lion [Leo]> begins setting in the morning and the Lyre [Lyra] sets.
○ ○
○ The Bird [Cygnus] begins setting acronychally.
○ ○ ○ ○ ○ ○ ○ ○ ○
○ Andromeda begins to rise in the morning.

Figure 2.1 Stone *parapēgma* fragment found at Miletus, published in Diels and Rehm (1904) as 456B and dated to the late second century BCE.

Source: Staatliche Museen zu Berlin-Preußischer Kulturbesitz Antikensammlung.

○○
○ The middle of the Water-Pourer [Aquarius] rising.
○ The Horse [Pegasus] begins to rise in the morning.
  ○
○ The whole Centaur [Centaurus] sets in the morning.
○ The whole Water-Snake [Hydra] sets in the morning.
○ The Sea Monster [Cetus] begins to set acronychally.
○ The Arrow [Sagitta] sets; the season of continuous
    westerly winds.
  ○○○○
○ The whole Bird sets acronychally.
○ <Arcturus rises> acronychally.[28]

(Most of the inscription survives, so very little has been inferred, indicated by the text provided here (by Diels and Rehm) in angled brackets; modern constellation names are in square brackets.)

The two stone *parapēgmata* found at Miletus are somewhat different from each other. That quoted above (which is dated to 109/8 and thought by Diels and Rehm to be approximately two decades older

Figure 2.2 Stone *parapēgma* fragment found at Miletus, published in Diels and Rehm (1904) as 456A and dated to the early first century BCE.

Source: Staatliche Museen zu Berlin-Preußischer Kulturbesitz Antikensammlung.

than the other, which is undated) may have simply listed astronomical events with almost no associated weather phenomena; only the west winds are mentioned. (But it should be noted that this is a fragment and so there is no way of knowing what was in the missing text.[29]) To judge from the surviving fragments, the other *parapēgma* appears to have provided a more extensive correlation of weather to celestial phenomena.

Further, in the later *parapēgma*, as is common in the literary forms of *parapēgmata*, various authorities are named, including Euctemon, Eudoxus, Philippos, 'the Egyptians' and 'the Indian Callaneus' (who is otherwise unknown). References, both in stone and in literary *parapēgmata*, and in other ancient writings, indicate that a good number of well-known ancient astronomers, including Euctemon and Hipparchus, produced their own astrometeorological *parapēgmata*. For the most part, these do not survive, but some scholars have attempted to 'reconstruct' these *parapēgmata* from the references and information that does survive in others.[30]

Here is an example of the text from the left-hand column of the second Milesian *parapēgma* (456A):

○ The Pleiades set in the evening according to Eudoxus,
 but according to Callaneus the Indian . . .
○ The Pleiades set in the evening and indicate hail
 [○] ○ ○ ○ [first day inferred by spacing]
○ The Hyades invisible in the evening, indicate hail and
 westerly wind blowing according to Euctemon,
 but according to the Indian. . . .[31]

The format of the later *parapēgma*, with references to earlier authorities and with weather phenomena correlated to stellar risings and settings, is typical of the literary *parapēgmata* which survive.[32] While the intended audience for each type of *parapēgmata* is not entirely clear, it is likely that the inscribed stone tablets were displayed in public, while the written texts were intended for the private use of individuals. Further, there are questions about where the stone *parapēgmata* would have been displayed; given the cost of erecting the inscribed *parapēgmata*, it seems likely that it would have been in an official space. And, because of the importance of keeping the peg in the correct place in the *parapēgma*, it seems very likely that the responsibility for moving the peg, on a daily basis, would have been an official duty.[33]

In understanding the *parapēgmata*, the question could be asked: 'What would have been the use of *reading* what would have been visible to the naked eye?' The *parapēgmata* eliminated the need to look at the heavens, to make astronomical observations to predict the weather; instead, a *parapēgma* could be consulted to learn what was meant to be occurring in the sky. Rather than having to rely on a clear sky and one's own ability to identify the right constellation at the proper time, the *parapēgmata* offered information and predictions collected and presented by others, many of whom (as we shall see) were 'known' experts. (Here, the *parapēgmata* are different from Hesiod's *Works and Days*, in which the reader is actually expected to look for and recognize specific astronomical events, and to correlate those events with Hesiod's text.[34])

It is impossible to know and difficult to guess how many stone *parapēgmata* were made and displayed. Very few examples have been found and it is not always clear that those stone fragments that some scholars have identified as *parapēgmata* actually have much in common with the examples from Miletus. The two *parapēgmata* from Miletus are themselves rather different from each other. Of the other inscriptions which have been identified as *parapēgmata*, the example found at Puteoli appears to be closely related to the second

Miletus type, seeming to relate weather to astronomical events.[35] It should also be noted that those examples which are the best preserved may not be representative of what constituted the entire genre; it is possible that they are somewhat idiosyncratic.[36]

What is apparently the earliest written example of a Greek *parapēgma* is contained in the Hibeh collection of papyri fragments found in Egypt on the upper Nile. The fragment comes from the cartonnage (or casing) of two mummies found by B.P. Grenfell and A.S. Hunt in 1902 and 1903. The *parapēgma,* from the Saïte nome (or district),[37] is part of a longer didactic work presented in epistolary format and was probably written at the beginning of the third century BCE.[38] The introduction does not survive in its entirety, but appears to have been an elementary overview. The *parapēgma* that follows the introduction utilizes the Egyptian calendar, and contains information about astronomical events, winds, festivals, the appearance of birds and the rising of the river, but not much about the weather. The author, who mentions that he has lived in Sais for five years, indicates in the fragmentary introduction that the source of his information is 'a wise man and a friend'; Otto Neugebauer suggested that this friendly reference was not merely a rhetorical device, and may have been true.[39]

Another important literary example is the one often referred to as the 'Geminus *parapēgma*'. The manuscript copies of Geminus' *Introduction to the Phenomena (Isagoge)* are followed without any interruption by a text of this *parapēgma;*[40] this had led some scholars to assume that the *parapēgma* is the work of Geminus himself (first half, first century BCE). But this attribution has been questioned and a much earlier dating has been suggested, to the second century BCE.[41]

The *parapēgma* appears under the title 'The time that the sun takes to travel through each zodiac sign and, for each sign, traditional predictions'.[42] This is a sample of the information provided:

> The Sun passes through the Water-Pourer [Aquarius] in 30 days.
> And then, on the 2nd, according to Callippus, the Lion [Leo] begins to set. Rain.
> And on the 3rd according to Euctemon, the Lyre [Lyra] sets in the evening. Rain. According to Democritus, stormy.
> And on the 4th day, according to Eudoxus, the Dolphin [Delphinus] sets acronychally.

And on the 5th day, according to Eudoxus, the Lyre sets acronychally. Rain.[43]

The 'Geminus *parapēgma*' is generally regarded as a compilation of information from earlier *parapēgmata*; the same has been argued for the later Milesian stone *parapēgma*. In each case, it is not clear whether the 'author/compiler' compiled the information from earlier authorities himself, or copied the compilation complete from another text.[44] Such compilations were not unknown in antiquity; indeed, some of the most important Greek scientific texts, including the Euclidean *Elements* and the Hippocratic corpus, are generally regarded as compilations.[45] But there is an interesting difference here: the author/compiler of the 'Geminus *parapēgma*' cites others as authorities (and probably as sources). The earliest authority cited is Meton (fl. 430 BCE) who is mentioned only once,[46] but his contemporary Euctemon (fl. 430) is named forty-seven times. Democritus (fl. 430 BCE), who Diogenes Laertius (9.48) reports produced his own *parapēgma*, receives eleven citations, while Eudoxus (fl. 370) is mentioned sixty times; Callippus (fl. 330) is named thirty-three times and Dositheus (fl. 230; a pupil of Konon) four times.[47] Euctemon and Eudoxus are both mentioned in the second Milesian *parapēgma*, which names other astronomers (e.g. the otherwise unknown Indian Callaneus) not cited in the 'Geminus *parapēgma*'.

This is one of the interesting features of the *parapēgmata*; in most surviving examples earlier authorities (including some of the great names of ancient astronomy) are cited. The *parapēgmata* have a tradition-laden format, which builds on the contributions of named predecessors. But, while they were presented as part of a tradition, the *parapēgmata* (presumably) were not simply records of past observations of astronomical and meteorological phenomena, but were intended to guide weather predictions.[48] Yet the tradition-bound format of the *parapēgmata* presented potential difficulties for their use as predictive tools; these difficulties will be discussed below. In addition, the form of the *parapēgmata* raises questions. What was the status of the authorities? Why were they cited? Before we return to these questions, some other related texts should be considered.

There are other works that appear to relate to the *parapēgma* tradition. In attempting to reconstruct a history of this tradition, some later sources are helpful, but often these provide only partial information. In the case of one important work on weather prediction, it is impossible to establish a firm attribution or date. This work, known as *On Weather Signs*, often attributed to Theophrastus (late fourth/early third century BCE),[49] indicates that the author of the text was at least aware of the tradition of correlating astronomical phenomena and weather. Patrick Cronin has suggested that it was written by a member of the Lyceum (neither Aristotle nor Theophrastus) around 300 BCE, possibly a student of Theophrastus.[50] Emphasizing the character of *On Weather Signs* as a collection of signs presented as a list, David Sider also subscribes to the view that the work has the stamp of the Peripatetic school. He has noted that, as a list, the work has something in common with the other lists attributed to Aristotle by Diogenes Laertius, and has suggested that *On Weather Signs* may have been collected by a member of the Peripatos and used by Aristotle (and/or by Theophrastus) in a similar way to the collection of city-state constitutions commissioned prior to the writing of the *Politics*.[51] Or, it is possible that *On Weather Signs* may have been an abridgement of a work on weather by either Aristotle or Theophrastus. Cronin has argued that the anonymous author utilized two other written sources (possibly a work on signs by Aristotle, now lost, as well as a work *On Signs* by Theophrastus, also now lost).[52] In any case, it seems likely that the work is an early product of the Peripatetic school; it is worth mentioning that it is always found in manuscripts containing other Peripatetic works. If this loose dating is correct, *On Weather Signs* predates those examples of *parapēgmata* which survive.

The author presents a variety of signs of weather. For example, as signs of rain the author offers: 'If there is a cloud in front of the setting sun which splits up its rays, this is a sign of stormy weather. And if the sun sets or rises while burning hot and there is no wind, this is a sign of rain.'[53] As Sider notes, *On Weather Signs* is largely a collection of signs presented as a list of conditional sentences: '"If X, then Y", where X is the sign and Y is the outcome: wind, rain, storm, or fair weather.'[54] Cronin emphasizes the author's reliance on popular weather lore,[55] and suggests that the predictive phenomena employed in the work may be grouped into the following categories: phenomena of sunrise, sunset and lunar phases; other celestial phenomena; domestic, marine, pathological (reactions of the human body), seasonal, zoological and botanical phenomena.[56] Here, for the most part, only celestial signs will be considered.

The work opens with this statement: 'We wrote down the signs of rain, winds, storms, and fair weather as follows to the extent that we were able, some of which we ourselves observed, others we took from others, men of no small repute. Those signs, then, that have to do with the stars as they set and rise must be learned from the astronomers.'[57] The author/compiler of On Weather Signs was directing his readers to experts and their technical works for information on weather signs associated with stellar risings and settings, just the sort of information which was contained in the parapēgmata.

From the very beginning of On Weather Signs, two different types of sources of information are pointed to as valuable: first-hand observations and the reports of recognized authorities; further, in some cases, it is clear that specialist knowledge is required. There is an ambiguous hint that the compilation represents a group effort. As Sider has noted, at the very beginning of the text the author uses a plural verbal form 'we have written';[58] this may be intended to signal that the work which follows is based on the combined efforts of many, and to strengthen the weight of the anonymous authority behind it.

The author then explains (1.3) that there are other, non-astronomical, weather signs that may be observed and studied. Some weather signs are peculiar to particular sorts of terrain, which have their own weather conditions; he briefly outlines the effects of mountains on the production of rain in valleys. Accordingly, the reader is warned to consider carefully local conditions and is further advised that it is worthwhile consulting individuals who have acquired their weather signs on the basis of local knowledge and experience; their information will be the best. In other words, generalized statements will be of limited use; the specific location of an observer will play an important role in interpreting the phenomena and signs of weather.

The author then states (1.4) that 'some persons have become good Astronomers in particular places'. He lists some 'good Astronomers', specifically noting where they were from or where they did their work: Matriketas of Methymna (Lesbos) made his observations from Mount Lepetymnos, Kleostratos of Tenedos from Mount Ida, and Phaeinos at Athens (whose pupil Meton established the cycle of nineteen years) observed the phenomena of the solstices from Mount Lykabettos.[59] As Sider notes, the author has made an effort to gather information from a relatively large area.[60] The author (1.4) adds that others have done astronomy in similar fashion, conveying a sense that there were a number of astronomical practitioners, that astronomy was not an activity confined to a few. The location of these observers should be noted; obviously a mountain-top would provide a good vantage point.[61]

It is characteristic of the parapēgmata that earlier 'authorities' are cited. Accordingly, the list of 'good Astronomers' in On Weather Signs

(1.4) is significant in that it suggests that the tradition of naming predecessors was fairly old.[62] By singling out particular astronomers by name, as well as place, the author emphasizes the role of individuals, while simultaneously making it clear that a larger number of people were engaged in astronomical activity, at least some of which was (apparently) useful for weather prediction. Bernard Goldstein and Alan Bowen, in a seminal article, suggested that the construction of *parapēgmata* was a defining characteristic of early Greek astronomy;[63] weather prediction may have been a motivating factor in much early astronomical work.

An important, and much later, *parapēgma* text which should be considered here is that known as the *Phases of the Fixed Stars and Collection of Weather Signs*, by the second-century CE Alexandrian astronomer Claudius Ptolemy. Ptolemy's *Phases* is an important astro-meteorological text that includes a *parapēgma*. While only the second book of the *Phases* has survived (and it is not certain what was contained in the first), the work is clearly part of the traditional genre of texts which correlate astronomical phenomena to weather, and it follows the traditional format, while introducing a few innovations.[64] The text is also an important source of information about the *parapēgma* tradition, providing information about Ptolemy's predecessors. In his *parapēgma* Ptolemy names a number of them and their observations and predictions are listed. The following is a sample of the text for the month of Choiak (corresponding to the end of November):

[Choiak]
1. [at *klima* (i.e. inclination, referring to a zone or band of terrestrial latitude which shares phenomena, including the length of longest daylight)][65]
    14½ hours: The Dog sets in the morning.
    [at] 15 hours: The bright star in Perseus sets in the morning. According to the Egyptians, south wind and rain. According to Eudoxus, unsettled weather. According to Dositheus, the day is ominous [meaning, the weather may change].[66] According to Democritus, the sky is turbulent, and the sea generally also.

    . . .
5. [at] 15½ hours: The star in the western shoulder of Orion rises in the evening. According to Caesar, Euctemon, and Eudoxus and Callippus, it is stormy.[67]

Looking at the list of authorities named by Ptolemy in the *Phases*, we see that some of them appeared either in the 'Geminus *parapēgma*' or in the Milesian stone inscription, including Meton, Euctemon, Democritus, Eudoxus, Philippus, Callippus and Dositheus and an anonymous group of 'Egyptians' (also mentioned in the second Milesian *parapēgma*).[68] But some 'new' names appear as well, and not all of them belong to astronomers who post-date Dositheus.[69] Ptolemy mentions Metrodorus (it is not clear if this is Metrodorus of Chios, fourth century BCE), Konon (first half of third century BCE, teacher of Dositheus), Hipparchus (190–120 BCE) and Caesar (here possibly referring to the work of Sosigenes, *c.* 50 BCE).[70] How did one qualify to be listed in Ptolemy's *parapēgma*? Did those named have good reputations as observers? Were they known as 'good Astronomers' (to use the phrase in *On Weather Signs*)?

The usefulness, and attendant valuing, of the knowledge and skills of specialist astronomers may explain why individuals are named in *parapēgmata*. Of course, in naming one's sources, the author has an opportunity to demonstrate how learned he is. But, generally, surviving astronomical works (including those of Ptolemy) tend not to name and discuss the work of predecessors, with some few exceptions. In fact, in some other genres, including some philosophical texts, predecessors are named only so that their ideas or methods can be argued against or disqualified.[71] So, the naming of predecessors as authorities is one of the distinguishing features of the *parapēgma* tradition. By naming the individual the parapegmatist asserts the reliability and specialist source of his information: this confirms the value of the tradition in asserting the possibility of useful prediction. The future may be uncertain, but predictions are made against a background of a tried and time-tested method. Astrometeorology and the *parapēgmata* are part of a scientific tradition in which tradition itself is valued.

One of the potential problems in using a *parapēgma* is that the observers cited were operating at different times and at different places, yet the *parapēgmata* do not, for the most part, indicate this. Did the compilers of the so-called Geminus and the Milesian *parapēgmata* assume that local conditions and observations could be generalized? Or did they trust that the astronomers listed by name were so well known that their places of operation did not need to be stated? Other factors may have been influential. It is, for example, possible that the work of Athenian astronomers, such as Meton and Euctemon, would have been valued in Miletus, regarded as a daughter-city of Athens.[72] Further, it must be significant that the stone *parapēgmata* were stationary: their location would have been obvious to users. As we have seen, specific astronomers (and their observations and predictions) were named; the location at which their astronomical work was done might have been inferred. But it is possible that the implications of utilizing data produced in other places and at other times were overlooked by many who would have seen the *parapēgmata*. Although, as we shall see, some did comment on the issues involved, that the predictions were made at specific (and different) places may not have mattered a great deal to all users of the *parapēgmata*. The list of ancient authorities lent an air of trustworthiness to the entire enterprise. Some sort of 'uniformitarian' principle was assumed (that is, that natural events occur in the same way, and with the same intensity, regardless of the time in which they happen). So, astronomical information from the distant past could still be useful in making new weather predictions. But, there are some authors, for example Theophrastus in his treatise *On Winds*, who suggest that climatic conditions have changed over time.[73] This view of the instability of weather and climate patterns would weigh against the longevity of the astrometeorological tradition. Nevertheless, it is clear that astrometeorological *parapēgmata* that listed earlier observations and predictions were produced over a long period of time. But, as will be discussed below, both a non-specialist author, Pliny the Elder, and the celebrated astronomer Ptolemy emphasized that knowledge of the location in which the astronomer worked was crucial.

Moreover, the role of the stone *parapēgmata* needs to be compared to that of other types of inscriptions. Rosalind Thomas has pointed out that 'little work has been devoted to the precise role of inscriptions, which is indeed usually taken for granted'. She has argued that 'the details of the inscriptions themselves, and the way the Greek writers treat them, make it clear that they were often thought of primarily as symbolic memorials ... rather than simply documents'.[74] But it seems unlikely that the *parapēgmata* were merely symbolic. Edwin A. Judge, in considering the rhetoric of inscriptions, has emphasized their public nature: by their display, they were intended to extend their message far into the future.[75] Yet, the usefulness for future users of astronomical observations and weather predictions made in a specific (possibly different) place at a specific (and possibly distant) time in the past is not clear; indeed, some ancient authors expressed that concern.

Pliny the Elder complained about the difficulties caused by various authors having made their observations at different locations, explicitly stating, at length, that observations and predictions cannot be generalized from place to place.[76] Pliny acknowledges that specialist astronomical knowledge and expertise are required for making astrometeorological predictions and asserts that such predictions are worth making, even though he recognizes that they may not be fulfilled, due to other influences and effects on the weather.[77] Others, including Geminus and Ptolemy, emphasized that knowledge of the location in which the astronomer worked was crucial. We do not know the relationship between the so-called 'Geminus *parapēgma*' and Geminus' *Introduction to the Phenomena*, but the *Introduction* appears to have been written much later. In it, Geminus pointed to the problem of generalizing from place to place when using a *parapēgma*. He explained (17.19) that the same *parapēgma* cannot be valid for Rome, Pontus, Rhodes and Alexandria.

These cautions make the assertion of the possibility of weather prediction, following careful and skilful observation, all the more striking. Clearly there was a strongly held view that weather phenomena could be reckoned with and possibly even understood; the *parapēgmata*, both literary and inscribed, bear witness to this. Writing of other types of stone inscriptions, Thomas has suggested that 'the inscription is a monument or memorial whose public presence and very existence guarantee the continuing force' of the document, often a legal instrument, which is inscribed.[78] In addition to serving as a means to track and correlate astronomical and weather events, it may be that the stone *parapēgmata* signalled to viewers that weather prediction was possible and rational and not simply subject to arbitrariness and capriciousness. (These latter behaviours had been associated with the traditional gods who, in some Greek writings, were described as controlling weather phenomena for their own purposes.[79]) Further, an astronomical calendrical tool linked to natural phenomena may have had a certain power and appeal not held by the calendars of individual city-states.

But even though weather prediction was possible, Pliny made it clear that care must be taken in utilizing the observations made by others, especially with regard to their locations.[80] Ptolemy, certainly more of a specialist astronomer, also addressed the problem of relying on the observations of others, citing (for example, in the *Mathematical Syntaxis*, also known as the *Almagest*) Hipparchus' distrust of Timocharis' astronomical observations taken at Alexandria.[81] And in the *Phases*, following the *parapēgma* itself, Ptolemy listed the places of observation that he assumed for each observer and the latitude (*klima*) for which he thought their results should be valid.[82] Ptolemy stated his concerns regarding the limitations of generalizing information from specific locations; clearly he believed that it is necessary to have information for each locality.[83]

Yet, in spite of Ptolemy's reservations about the reliability of his predecessors' observations and the limitations of observations made in a particular locality, it seems that he valued the tradition and approach of the *parapēgmata*. After all, the *Phases* comprised two books in which Ptolemy not only presented his own *parapēgma*, but also introduced certain important innovations (e.g. working with individual stars of the first magnitude, rather than constellations). It is difficult to imagine Ptolemy expending the effort necessary for such a project if he were as sceptical about the *parapēgmata* tradition as some historians, for example Otto Neugebauer, have suggested.[84] Of course, it is possible that Ptolemy was working for a patron, perhaps the Syrus to whom some of his other works were addressed, who may have requested such a work. Or, 'Syrus' may have simply have been a fictional dedicatee. In any case, it should be acknowledged that the astrometeorology of the *Phases* fits well with other parts of Ptolemy's work.

It is important to note Ptolemy's own strong interest in prediction (of weather and of astronomical phenomena), and his acknowledgement that there are limits to what can be achieved. The desire to be able to make predictions, in spite of recognized problems and limitations, was very compelling. In a number of texts Ptolemy affirmed that it is possible to predict certain events.[85] He recognized that the determination of fixed-star phases was itself problematic and declined to compute them, noting the difficulties. He was something of a pragmatist, content to achieve the possible, relying on what was available. In the *Syntaxis*, he explained that: 'For the time being we content ourselves with the approximate [phases] which can be derived either from earlier records or from actual manipulation of the [star]-globe for any particular star.' Here he also acknowledged that weather predictions based on stellar risings and settings are almost always approximations; he suggested that it is not the actual times of first and last visibility, but rather the configurations, which are important causally.[86]

Otto Neugebauer, in *A History of Ancient Mathematical Astronomy*, gave an indication of the use of astronomical knowledge for weather prediction. He noted that 'the Almagest contains several sections which were only added by Ptolemy in order not to omit from his works topics which traditionally belonged to mathematical astronomy'. As an example, he points to Book 6, where Ptolemy ends his discussion of lunar theory with a chapter (6.11) which is of little astronomical significance, dealing with 'the determination of certain angles which were considered, apparently by a very old tradition, of importance for weather prognostication'. Neugebauer explained that the moon's sickle or latitude were 'considered significant for storms and weather according to the part of the horizon toward which they are "inclined" since the segments of the horizon are naturally associated with the winds and the weather they bring'.[87] Ptolemy discusses the angles of inclination at eclipses at some length, without (in the *Syntaxis*) detailing their usefulness, though he briefly mentioned the observation of eclipse-related (weather) indications (τὰς ἐκλειπτικὰς ἐπισημασίας) earlier in Book 6.[88] In his discussion of the determination of these angles of inclination, Ptolemy stresses that accuracy is not critical. He notes that a rough estimate of the angles is sufficient and aims to provide a convenient method for achieving the necessary results.[89] Neugebauer is persuasive in his argument that this treatment of inclination is not astronomically motivated; Ptolemy's desire to provide a simple way to estimate the angles indicates that he was hoping to appeal to those who wanted to use celestial events to predict weather, without elaborate astronomical techniques.[90] As an accomplished astronomer, Ptolemy recognized that he had something to offer to other practitioners who were not as technically and theoretically adept. In other works as well, Ptolemy made it clear that he regarded it as part of his own contribution to provide easier means of making astronomical determinations and calculations.[91]

The important relationship between atmospheric phenomena and celestial bodies is also emphasized in the *Tetrabiblos* (2.11–13), where Ptolemy detailed some of the signs useful for weather prediction. At the beginning of the *Tetrabiblos*, which he indicated may be regarded as a companion volume to his *Mathematical Syntaxis*, Ptolemy addressed the means of 'prediction through astronomy': he explained that there are two which are 'most important and valid'. The first type, which investigates celestial motions, had already been covered in the *Syntaxis*; the second, which studies celestial influences, was the subject of the *Tetrabiblos* itself.[92]

While he acknowledged that the second type of astronomy was less exact than the first, Ptolemy nevertheless held that it is 'evident that most events of a general nature draw their causes from the enveloping heavens';[93] for this reason it is desirable to study both types of astronomy. He went on (*Tetrabiblos* 2.12) to elaborate the ways in which the heavenly bodies and their motions affect terrestrial events, and specifically addressed the question of predicting seasonal changes and weather; in fact, he devoted a fair amount of space to discussing weather prognostication based on astronomy. He stated that 'in general we see that the more important consequences signified by the more obvious configurations of sun, moon, and stars are usually known beforehand, even by those who inquire, not by scientific means, but only by observation'. He explained that: 'Those which are consequent upon greater forces and simpler natural orders, such as the annual variations of the seasons and the winds, are comprehended by very ignorant men, nay even by some dumb animals; for the sun is in general responsible for these phenomena.'[94]

He explained that some predictions are made as a result of special experience, so, for instance, 'sailors know the special signs of storms and winds that arise periodically by reason of the aspects of the moon and fixed stars to the sun'.[95] But Ptolemy made it clear that only specialist astronomers would have sufficient information and understanding to make the best weather predictions:

> If, then, a man knows accurately the movements of all the
> stars, the sun, and the moon, so that neither the place nor
> the time of any of their configurations escapes his notice, and
> if he has distinguished in general their natures as the result
> of previous continued study, even though he may discern,
> not their essential, but only their potentially effective qualities, such as the sun's heating and the moon's moistening,
> and so on with the rest; and if he is capable of determining
> in view of all these data, both scientifically and by successful conjecture, the distinctive mark of quality resulting from
> the combination of all the factors, what is to prevent him
> from being able to tell on each given occasion the characteristics of the air from the relations of the phenomena at the
> time, for instance, that it will be warmer or wetter?[96]

Ancient meteorologists, as Ptolemy knew, used various means of prediction, including celestial signs. However, he clearly believed that astronomers make the best weather forecasters.

In the second book of the *Tetrabiblos* (2.13), Ptolemy considered the use of observations of astronomical events in predicting changes in the weather; he explained that:

> Observations of the signs that are to be seen around the sun,
> moon, and planets would also be useful for a foreknowledge
> of the particular events signified [or, 'the changes in the

weather' = *episēmasiai*]. We must, then, observe the sun at
rising to determine the weather by day and at setting for the
weather at night, and its aspects to the moon for weather
conditions of longer extent, on the assumption that each
aspect, in general, foretells the condition up to the next.[97]

He goes on to give some details of the sorts of observations to be
made: for example, 'when the sun rises or sets clear, unobscured,
steady, and unclouded, it signifies fair weather'. But, 'if at rising or
setting it is dark or livid, being accompanied by clouds, or if it has
haloes about it on one side, or the parheliac clouds (παρήλια νέφη)
on both sides, and gives forth either livid or dusky rays, it signifies
storms and rain'.[98] Weather indications based on the appearance of
the moon and of the fixed stars are also included, as are various 'occasional phenomena in the upper atmosphere' (τῶν ἐπιγινομένων δὲ
κατὰ καιροὺς ἐν τοῖς μετεώροις),[99] which foretell droughts or winds,
while shooting stars may indicate various types of winds and storms,
depending on the angle (γωνία) from which the star itself comes.[100]
Rainbows signify storms after clear weather and clear weather after
storms, while clouds resembling flocks of sheep sometimes indicate
storms.[101] (Notice that these indications of weather are based on
actual observation of the sky, rather than on the consultation of information regarding stellar risings and settings set out in a *parapēgma*.)

Ptolemy (*Tetrabiblos* 2.10) discussed the general usefulness of
astronomy/astrology for predicting the weather, and for knowledge
of the seasons. He explained that 'the sun creates the general qualities and conditions of the seasons, by means of which even those who
are totally ignorant of astrology can foretell the future'. He went on
to advocate that 'we must take into consideration the special qualities
of the signs of the zodiac to obtain prognostications of the winds and
of the more general natures'.[102] Ptolemy then (2.11) elaborates the
nature of the signs of the zodiac, and their weather effects, beginning
with Aries, which is characterized by thunder or hail, through the
year to Pisces, which is, on the whole, cold and windy. He had earlier
(1.4) explained the powers (δυνάμεις) of the planets to heat, cool, dry
and humidify. The powers of the fixed stars are detailed in chapter 9
of Book 2, where the various powers are discussed with reference to
those of the planets.

The belief that astronomical knowledge could be useful for
weather prediction was a consequence of Ptolemy's acceptance of a
long-standing tradition, which posited that the heavenly motions
in some way cause most events of a general nature on earth.[103]

By understanding the motions of the heavenly bodies, one will be able to gain important insights about events on earth. The vision of a cosmos in which celestial motions have important influence on terrestrial events owes much to the idea of the harmonious fitting together of the universe; the harmony of the heavens is crucially linked to earthly events.[104] This linking together of the celestial and terrestrial into a harmonious universe made possible one tradition of predicting meteorological events. For some ancient authors, astrometeorology depended quite literally upon the harmony of the cosmos.

Weather signs

There is a question that should be raised regarding the relationship between the 'signs' of weather and the 'causes' of weather. Did those who compiled and referred to the astrometeorological *parapēgmata* regard the association of particular risings and settings with specific weather conditions as coincidental or correlated? Was a causal link suggested between correlated phenomena? Certainly, some authors regarded the connection between the celestial and the meteorological as causal. For example, Pliny (*NH* 2.39.105) clearly accepts the idea that celestial events do have seasonal and meteorological influence. He asks the question 'who can doubt that summer and winter and the yearly vicissitudes observed in the seasons are caused by the motion of the heavenly bodies?' He elaborates, explaining that 'as the nature of the sun is understood to control the year's seasons, so each of the other stars also has a force of its own that creates effects corresponding to its particular nature'. He provides examples, and goes into some detail, pointing to the rainstorms associated with Saturn and the heat accompanying the rising of the Dog-star.[105]

But other writers indicate that this view of the celestial causes of atmospheric phenomena was not universally held and, indeed, was criticized by some. Geminus acknowledges that it was generally held that Sirius causes the heat of the 'dog-days' of summer, but he argues that those people who hold that view are mistaken, since the star only serves to mark the season when the heat of the sun is greatest.[106] Epicurus (341–271 BCE) famously argued that multiple possible causes, rather than a single explanation, should be considered for natural phenomena. He suggested that 'the signs in the sky which betoken the weather may be due to mere coincidence of the seasons, as is the case with signs from animals seen on earth, or they may be caused by changes and alterations in the air'. He concluded that 'neither the one explanation nor the other is in conflict with facts, and it is not easy to see in which cases the effect is due to one cause or to the other'.[107] While Epicurus was keen to promote his own programme of multiple explanations, and so was unwilling to endorse a single explanation regarding weather signs, his discussion provides further evidence that both views were held.

Some modern writers tend to portray ancient science as mainly speculative and theoretical, with little observation involved. In fact, the evidence of the astrometeorological *parapēgmata* suggests that observations of natural phenomena were undertaken and taken seriously. Further, as some of the texts we have looked at indicate, the proper methods for observing and evaluating others' observations were considered and discussed, not only by those engaged in the activity. Some critics recognized that astrometeorology was not without value, even if certain practitioners (often referred to as 'Chaldeans') used suspect methods.[108] Columella (fl. 50 CE) in his work *On Agriculture* (Book 11.1.30–2), explained that:

> warning about the duties of each month, dependent on a consideration of the stars and sky, is necessary: for as Vergil [*Georgics* 1.204ff.] says:
>> Arcturus' star, the Kids and gleaming Snake
>> We must observe as carefully as men
>> Who, sailing homewards o'er the wind-swept sea
>> Through Pontus and Abydos' narrow jaws,
>> The breeding-ground of oysters, seek to pass.

After quoting the poet as an authority, Columella goes on to say that:

> Against this observation I do not deny that I have disputed with many arguments in the books which I wrote *Against the Astronomers* [now lost]. But in those discussions the point which was being examined was the impudent assertion of the Chaldaeans that changes in the air coincide with fixed dates, as if they were confined within certain bounds; but in our science of agriculture [*in hac autem ruris disciplina*] scrupulous exactitude of that kind is not required, but the prognostication of future weather by homely mother-wit, as they say, will prove as useful as you can desire to a bailiff, if he has persuaded himself that the influence of a star makes itself felt sometimes before, sometimes after, and sometimes on the actual day fixed for its rising or setting. For he will exercise

sufficient foresight if he shall be in a position to take measures
against suspected weather many days beforehand.[109]

In other words, farmers would benefit from being able to plan and
act ahead of weather changes, even if the time of those changes could
not be precisely predicted. It is also interesting to note that Columella
very clearly distinguished between the ways in which the 'rustic' and
the 'astrologer' marked the seasons (Book 11.2.2): 'the husbandman
ought not to observe the beginning of spring, in the same way as
the astronomer, by waiting for the fixed day which is said to mark the
entry of spring, but let him even take in something also from the
part of the year which belongs to winter, since, when the shortest
day is passed, the year is already beginning to grow warmer and
the more clement weather allows him to put work in hand'.[110]
Clearly, according to Columella, the farmer has rather different
concerns from the astronomer, and so need not abide by a fixed
astronomical calendar. Here, the suggestion is that, while astronomically based almanacs may be useful for farmers, the good husbandman cannot rely on them blindly. It should be noted that
Columella's detailed agricultural almanac, presented in Book 11, does
not cite or name astronomers, though stellar risings and settings
form the basis.[111]

Another first century CE Roman author with agricultural interests,
Pliny the Elder, relates (*NH* 18.60.225) that clothing merchants
relied on astronomical predictions, noting that retailers used a study
of the stars to help set prices for the coming season. He explained
that 'dealers out to make money, who are careful to watch for chances,
make forecasts as to the winter from its [the Pleiades'] setting: thus
by a cloudy setting it foretells a wet winter, and they at once raise
their prices for cloaks, whereas by a fine weather setting it foretells a
hard winter, and they screw up the prices of all other clothes'.[112]
So strong practical interests may have motivated some astrometeorological predictions.

Sextus Empiricus (end of the second century CE) (*Against the
Professors* = *Adversus Mathematicos* 5.1) contrasts the goals of the astro-
nomical techniques used for weather prediction by Eudoxus and
Hipparchus with what he considers to be the superstitious practices
and prophecies of human events indulged in by Chaldean astrologers.
Sextus' lively passage is worth quoting:

> The task before us is to inquire concerning astrology or the
> 'Mathematical Art' – not the complete Art as composed of

arithmetic and geometry . . .; nor yet that of prediction practised by Eudoxus and Hipparchus and men of their kind,
which some also call 'astronomy' (for this, like Agriculture
and Navigation, consists in the observation of phenomena,
from which it is possible to forecast droughts and rainstorms
and plagues and earthquakes and other changes in the surrounding vault of a similar character); it is rather the casting
of nativities, which the Chaldeans adorn with more high-
sounding titles, describing themselves as 'mathematicians'
and 'astrologers,' treating ordinary folk with insolence in
various ways, building a great bulwark of superstition
against us, and allowing us to do nothing according to right
reason.[113]

Clearly, in Sextus' view, the aims of Eudoxus and Hipparchus were
vastly different from those of the Chaldean astrologers. While he
argued against the claims of the Chaldeans, he tacitly endorsed the
work of the Greek astronomers in predicting the weather.

Today we place a high premium on 'newness' and innovation and
sometimes seem to think that this should characterize scientific work.
The practice of astrometeorology was a long-lived tradition, span-
ning over a thousand years. It is striking that Ptolemy, whose
astronomical methods were still being used in the sixteenth century,
himself placed a high value on the tradition of astrometeorology.
In his discussion of earlier *parapēgmata*, he never claimed to be inno-
vating and coming up with something entirely new. Rather, he was
refining an existing approach, working within an accepted tradition.

The ancient writers on weather prediction never claimed that their
predictions would be foolproof and absolutely reliable. On the contrary, they recognized and pointed to the limits of their knowledge
and ability. They aimed to provide useful weather predictions, even
though they knew their accuracy could not be guaranteed. These
ancient authors clearly recognized the fallibility of their approach.
They did not pretend to guarantee predictions or certain results.
Nevertheless, they were committed to the attempt to predict the
weather, within acknowledged limits. It is possible that, because
weather prediction was regarded as necessarily unreliable, the ability
to point to a long-standing tradition of weather prediction carried
great weight, and served to ground the practice as time-tested and
worthwhile.

Geminus, in his *Introduction to the Phenomena*, explained that the
weather predictions found in the Greek *parapēgmata* are based on

observations. These observations and predictions refer to the usual order of things, but may not always come to pass. He notes that an astronomer is not blamed if his weather predictions do not come true, but if his eclipse prediction is wrong, he will be rightly criticized, since it is based on a method which is more certain and precise.[114] Geminus devotes a fair amount of space (chapter 17) to denying that astronomical events cause weather phenomena. For Geminus, weather is not caused by celestial bodies, but celestial phenomena may serve as indications of meteorological conditions. He argues that the heat of the summer is not caused by the rising of the Dog-star, but by the warming of the sun; the Dog-star just happens to rise at the same time as the warmest part of the year.[115]

In spite of Geminus' arguments against the idea of causal relationships operating between celestial events and weather, it is interesting to note that the manuscript copies of Geminus' *Introduction* have appended a detailed *parapēgma*. It is difficult to know what to say about the relationship between Geminus' views on astrometeorology and the *parapēgma* which follows in the manuscripts.[116] Why was a *parapēgma* appended to Geminus' *Introduction*? Why would an argument against astrometeorology be circulated with a detailed astrometeorological text?[117] Is it an indication that Geminus' arguments were taken as just one side of the debate? It is not possible to answer these questions with any certitude, but even if one did not accept a causal relationship between the astronomical and the meteorological, the *parapēgma* could still have been useful for predictions.

There remain many questions about the ancient astrometeorological *parapēgmata*. In some cases we do not know who produced them or when. There is also the question, whom were they created for? The written versions were probably intended for private use, but by whom? It is possible that they were the stock-in-trade of professional astronomers, who were consulted by business people, including clothing merchants like those mentioned by Pliny, for advice about the coming season, much as today's commodity traders purchase specially commissioned meteorological forecasts to help them decide when to buy and sell pork bellies. That various named 'experts' were presented as providing the information may well have added to the appeal; possession of and access to the *parapēgma* could provide the appearance of a degree of mastery of weather prediction. We do not know the specific audiences for which the written astrometeorological *parapēgmata* were produced.

Turning to the stone *parapēgmata*, certain questions arise as to their intended and actual use. The provision of holes for the peg-marker suggest that they were actually meant to be used. Given the expense of producing an engraved stone object, it seems reasonable to assume that such *parapēgmata* were displayed in public. But how many people were literate and able to read them? William V. Harris has noted that 'there has never been a determined attempt to collect and analyse the evidence that does exist [of ancient literacy], except for Athens and Sparta in the classical period and for Graeco-Roman Egypt'. The level of literacy in Miletus in the relevant period is difficult to guess. For the classical period, Harris suggests that the difference in literacy between Athens and Miletus was probably not very great. As in Athens, ostracism (in which fragments of pottery were inscribed or painted as voting ballots) was practised in Miletus. But, as Harris explains, ostracism may only be evidence of the semi-literacy of male citizens (who were entitled to vote), for they needed only to be able to write the name of the person to be ostracized on the potsherd. Furthermore, it is possible that citizens were persuaded to use *ostraka* that were 'mass produced' with the appropriate name.[118]

For the Hellenistic period, Harris points to evidence of teaching in Miletus: 'at Miletus in 200 or 199 a citizen named Eudemus presented to the city a sum which was to produce 480 drachmas a year for each of four teachers (*grammatodidaskaloi*) in addition to some money for athletic trainers (*paidotribai*)'.[119] The inscription which records this donation shows that Miletus already had, prior to this, an education law (*paidonomikos nomos*) which regulated student and teacher 'displays'. This law indicates that the city was interested in the effectiveness of the schooling provided to pupils. But Harris argues that 'four teachers could not by themselves have educated all the sons of the free inhabitants of Miletus'.[120] Nevertheless, on this evidence, it seems reasonable to assume that many Milesian men, and perhaps even some boys, would have been able to read the Milesian *parapēgmata*. There is also evidence concerning literacy in Puteoli, where a fragment of another meteorological *parapēgma* was found. In Puteoli, when the collateral of defaulting debtors was sold, notice of the sales was announced in *libelli* posted in a public portico.[121] As an example of education, Harris also cites Pliny (*NH* 9.25) who mentions the son of a *pauper* attending a school (*ludus litterarius*), at Puteoli under Augustus.[122] So, for two of the cities where stone fragments of *parapēgmata* have been found, there is evidence of education and literacy during the relevant period, which suggests that the stone *parapēgmata* could have been read by at least some of the inhabitants. And, as Robert Hannah has argued, 'even if only a few could read [a *parapēgma*] and then understand it, nevertheless the intention would seem to have been that the parapegma should have at least a visual impact on more than just an élite, literate few', monumentalizing an activity (the use of an astronomical calendrical tool) important to the populace.[123] But if the *parapēgmata* were displayed in public, how were they protected from being tampered with by children and others having a lark? It would have been terribly easy, it seems, to move the peg to the wrong day and throw the whole community off schedule.

The astrometeorological *parapēgmata* have been discussed in some detail. To an extent, the almanacs presented in the works of the agricultural writers are related to these,[124] just as they are related, as well, to the final section of the Hesiodic poem, the *Works and Days*. Astronomical weather signs are described (in varying levels of detail) in these texts. The *Works and Days* is an important example of a poem that communicated astronomically linked seasonal weather information. Several of our best sources of information about weather signs (both astronomical and non-astronomical) are poems of a particular genre, known as didactic poems; these poems are also linked to the tradition of collecting weather information illustrated in *On Weather Signs*. For example, the *Phaenomena* written by Aratus of Soli (in Cilicia; *c.* 315 to before 240 BCE) and the *Georgics* of Virgil (70–19 BCE) are poems that contain astronomically based almanacs outlining seasonal changes and also detailed information of shorter-term weather signs.

The *parapēgmata* contain only astronomical signs of weather, and no non-astronomical signs, but it is important to recognize that in many other texts both astronomical and non-astronomical signs are described. In some cases, different types of weather signs are treated separately; in others (including the works by Aratus and Virgil) non-astronomical signs may be interspersed with astronomical indications. While the poetic format of some of these texts is significant, it should be emphasized that weather signs are also described in other genres and other textual formats. In turning to non-astronomical signs of weather, significant sources of information include, once again, the work known as *On Weather Signs* and also Pliny's *Natural History* (Books 2 and 18).

The work known as *On Weather Signs* is of special significance. Unlike the astrometeorological *parapēgmata*, it provides both astronomical and non-astronomical weather signs which are, for the most part, provided in a rather cursory format reminiscent of a list.[125] The use of a list-format features in other genres of meteorological texts as well, most notably that of didactic poetry. Presumably, such lists would have been relatively easy to memorize, and so offered a certain degree of usefulness. But as David Sider has pointed out, the arrangement of *On Weather Signs* appears to be somewhat impractical. Following the introductory remarks, the work is organized according to the various signs of a particular meteorological phenomenon. First, signs of rain are listed, then signs of winds, followed by storm, and then fair weather; a short series of signs linked to periods during the year, such as the equinox, close the work.

As Sider notes, the arrangement of each section 'goes its own way as it sets out in no particular order signs derived from meteorological activity . . ., animal behavior, plant growth, and some miscellaneous signs'. He suggests that 'if a farmer were interested in knowing whether rain was likely he could be imagined thumbing through this work to remind himself of what to look for this day. But . . . the most practical book of weather signs would arrange them first by sign and then by the event they signal'.[126] Thus, we can imagine that a handbook organized by the signs to be observed, such as 'fiery moon' or 'hawk on a tree', might have had a more obvious practical use.

The author of *On Weather Signs* gives an outline of the astronomical signs of weather; this discussion occupies the first part of the work. The author states his aim clearly in the opening line: namely, to describe the signs of rain, winds, storms and fair weather; each of these is dealt with separately. In each section, astronomical signs are presented first, followed by other weather signs. Finally, at the end of the work signs are described which affect the entire year, or large portions of it. Most of these signs are related to astronomical events (such as the spring equinox and phases of the moon) and phenomena (including comets). It is worth giving a sense of the text, not least because it seems to have served as a source for later authors, including Aratus, Virgil and Pliny.

Many different sorts of phenomena may serve as weather signs. So, for example, the author of *On Weather Signs* includes the following as signs of rain:

> [13] Shootings stars in great number are a sign of rain or wind which will come from the same direction as the shooting stars. And if at sunrise or sunset the rays set close together, this is a sign of storm. And when at sunrise its rays have the same color they do during an eclipse, this is a sign of rain, as well as when clouds resemble woolen fleece. Bubbles rising to the surface of rivers more than is usual are a sign of much rain. Often, an iridescent halo shining either around or through a lamp is a sign of rain from the south.

[14] If the winds are from the south, snuff on a lamp's wick signals rain; and in proportion to its amount it also signals wind, and if it is finely granulated like millet seeds and shiny it signals both rain and wind. It is a sign of rain when during winter it <the lamp> intermittently casts off the flame like bubbles. Likewise if the rays <of the flame> leap upon it, and if during winter snuff builds up.

[15] Nonaquatic birds bathing themselves are a sign of rain or storm. And a toad washing itself and frogs croaking more than usual signal rain. Likewise the appearance of the lizard which they call 'salamander'; and even more does a green frog singing in a tree signal rain. Swallows striking the surface of a lake with their bellies signal rain. An ox licking a forehoof signals storm or rain; if he arches his head up toward the sky and sniffs this signals rain. [127]

As can be seen from this passage (*On Weather Signs* 13–15), astronomical and atmospheric phenomena, as well as animal behaviours and other natural phenomena, may indicate specific weather conditions.[128] Many of these weather signs are directly echoed in other authors, including Aratus, Virgil and Pliny; some of these echoes will be noted below.[129] *On Weather Signs* may have been one of a larger number of texts devoted to detailing weather signs; ancient authors attributed at least two others (no longer extant) to Aristotle and Theophrastus.[130]

Among our main sources of information about signs of weather, some of the most significant are didactic poems. As Alessandro Schiesaro explains, didactic poetry was not regarded by Greek and Roman theorists as a separate genre, but modern scholars recognize didactic poems as those which were presented with the aim of teaching the reader about a particular subject. Such poems are usually nominally addressed to a specific individual, who may be taken as a model by readers. The subject-matter of didactic poems ranges from philosophy to hunting; other popular themes include farming and astronomy. Schiesaro makes the distinction between an older form of didactic poetry, in which general moral and philosophical teaching plays an important role (Hesiod's *Theogony* and the *Works and Days* would be examples of this early type) and a later type common in the Hellenistic period, characterized by subjects that may be specialized and obscure. Later didactic poems were designed to show poetic virtuosity; Aratus' *Phaenomena* is an example. Although written even later, Virgil's *Georgics*, seeming to focus on farming and touching on weather, clearly aims to address wider issues, and so may be seen as a latter-day parallel to Hesiod's poems, particularly the *Works and Days* (which was, after all, also concerned with agriculture).[131]

Having identified the subject-matter of didactic poetry, it is also necessary to consider the intended audiences for these poems. Alexander Dalzell noted that Servius (fourth century CE) in his commentary to Virgil's *Georgics* pointed out that 'a didactic poem must be addressed to someone; for instruction implies someone in the role of teacher and someone in the role of student'. Virgil's *Georgics* was addressed to his patron Gaius Maecenas; the *Works and Days* was addressed to Hesiod's brother Perses.[132] But not all didactic poems are addressed to specific individuals; Aratus' *Phaenomena*, an important astronomical poem which describes weather signs, was not so addressed. Nevertheless, the role of the implied addressee in didactic poetry is crucial, for it preserves the illusion of teaching and shared knowledge central to the genre.[133]

Aratus of Soli composed an enormously popular poem, known as the *Phaenomena*. The first part of the poem begins with what may be regarded as a dedication and hymn to Zeus:

> Let us begin with Zeus, the power we mortals never leave
> Unsaluted. Zeus fills all the city streets,
> All the nation's crowded marts; fills the watery deeps
> and havens; every labour needs the aid of Zeus.
> His children are we. He benignant
> Raises high signals, summoning man to toil,
> And warning him of life's demands: tells when the sod
>   is fittest
> For oxen and harrows; tells the auspicious hours
> For planting the sapling and casting every seed.
> 'Twas he who set the beacons in the sky,
> And grouped the stars, and formed the annual round
> Of constellations, to mark unerringly
> The days when labour is crowned with increase.
> Him therefore men propitiate first and last.
> Hail, father, mighty marvel! hail! mighty benefactor!
> Thyself and those who begot thee! And ye too, muses,
> Gracious influences, hail! and while I essay to tell of
>   skies
> What mortal may tell, guide right my wandering lay.[134]

Here, Aratus asserts the orderly arrangement of the heavens, credited to the beneficence of Zeus himself, which permits people to use the sky and its signs as the basis of an agricultural almanac; as Douglas Kidd has noted, the 'whole poem illustrates the presence of Zeus as sky and weather god and the providence of Zeus in providing the signs that are helpful to mortals'.[135]

Following the religiously and practically inspired proem, Aratus launches into a description of the constellations, and lists risings and settings of many of them, with the aim of telling the time at night; the practical value of the information provided is emphasized. In composing his poem, Aratus utilized a prose work by Eudoxus of Cnidos (c. 390–c. 340 BCE), known also as the *Phaenomena*, which contained detailed information about the constellations. The second part of Aratus' poem, given a separate title 'Signs' (= 'Διοσημεῖαι'), is concerned with signs of weather (and may be derived from the work mentioned earlier, known as *On Weather Signs*).[136]

Although he wrote other poems, Aratus' *Phaenomena* is his only extant work. The *Phaenomena* quickly became one of the most widely read poems in the ancient world, after the Homeric poems, the *Iliad* and the *Odyssey*. The popularity of Aratus' *Phaenomena* was widespread and long-lived. Various Roman writers translated it into Latin: Marcus Tullius Cicero (106–43 BCE), Germanicus Iulius Caesar (15 or 16 BCE–19 CE), and Postumius Rufius Festus Avienus (fl. mid-fourth century CE). It was one of only a very few Greek poems translated into Arabic. In antiquity, at least twenty-seven separate commentaries were written on the poem.[137]

Gerald Toomer has suggested that the appeal of the poem lay in its literary charm – including its many mythological allusions – rather than in its astronomical information. The poem was the subject of numerous ancient commentaries, some of which pointed out the astronomical errors committed by Aratus. For example, the sole surviving work of Hipparchus (who flourished during the second half of the second century BCE) is known as the *Commentary on the Phaenomena of Aratus and Eudoxus*. In this work, Hipparchus (whom Ptolemy was to call a 'lover of truth') criticized the descriptions of constellations given by both Aratus and Eudoxus. (Interestingly, Hipparchus' commentary was concerned only with the astronomical material in the poem, not with the weather signs.)[138]

But even if the astronomical information contained in the poem was not always accurate, its popularity ensured that the information conveyed there reached a wide audience. In any case, Aratus' poem is important for us, because it provides information about weather signs;[139] Peter Toohey has gone so far as to suggest that the primary focus of the poem, overall, is weather prediction: he claims that 'Aratus' poem is designed to allow one to predict the weather.' In Toohey's view, this interest is practically motivated: if the weather can be predicted it will then be possible to plan what sort of work and what kinds of activities can be safely pursued on particular occasions. For Aratus, astronomy provides the best means of weather prediction.[140] The first, astronomical, section of the poem (lines 1–732) sets the groundwork for the information about weather prediction which follows.

At line 732, Aratus notes that 'everywhere the gods give these many predictions to men'; the section on weather signs then follows: 'Don't you see? When the moon with slender horns is sighted in the west, she declares a waxing month.'[141] According to some interpreters, the role of the gods, specifically Zeus, is important throughout the poem as a whole. Peter Toohey has argued, persuasively, that the poem is 'virtually a hymn to Zeus'; he stresses that Zeus is the key to the poem:

> The *Phaenomena*, which gives us a practical and detailed designation of the usefulness of star-signs for daily life, provides a demonstration of the beneficence of Zeus. It illustrates Aratus' striking assertion [in the opening lines] that 'the streets are full of Zeus and all the market-places of men, and the sea is full of him and the heavens. Always do we all have need of Zeus'.[142]

On this reading, the order of the heavens provides a mirror of Zeus' will.[143]

Appropriately the poet then describes the phases of the moon, before providing a brief description of the zodiac. Aratus (lines 752–64) explains that:

> You too know all these (for by now the nineteen cycles of the shining sun are all celebrated together), all the constellations that night revolves from the belt to Orion again at the end of the year and Orion's fierce Dog, and the stars which, when sighted in Poseidon's realm or in that of Zeus himself, give clearly defined signs to men.
>
> Therefore take pains to learn them. And if ever you entrust yourself to a ship, be concerned to find out all the signs that are provided anywhere of stormy winds or a

hurricane at sea. The effort is slight, but enormous is later the benefit of being observant to the man who is always on his guard: in the first place he is safer himself, and also he can help another with his advice when a storm is rising near by.[144]

Aratus distinguishes between the cycles of seasonal weather and more irregular weather; he instructs his readers to be familiar both with the annual astronomical events, linked to seasonal changes, and with local signs which indicate impending changes.[145]

Aratus provides detailed instructions for discerning the weather signs provided by the moon (778–93):

Observe first the moon at her horns. Different evenings paint her with different light, and different shapes at different times horn the moon as soon as she is waxing, some on the third day, some on the fourth; from these you can learn about the month that has just begun. If slender and clear about the third day, she will bode fair weather; if slender and very red, wind; if the crescent is thickish, with blunted horns, having a feeble fourth-day light after the third day, either it is blurred by a southerly or because rain is in the offing. But if, when she brings the third day, the moon does not lean forward from the line of the two horn-tips, or shine inclining backwards, but instead the curve of the two horns is upright, westerly winds will blow after that night. But if she brings in the fourth day also similarly upright, she will certainly give warning of a gathering storm.[146]

It is worth noting that this section shares much with passages found in some of the other works considered here; for similar signs relating to the moon, compare *On Weather Signs* 51, 12 and 38, Virgil's *Georgics* 1.424–35 and Pliny's *Natural History* (18.79.347–50).[147]

Not all of the signs of weather were astronomical: 'Let a sign of wind be also a swelling sea and beaches roaring a long way off, sea-coasts reverberating in fair weather, and a mountain's summit-peak sounding.'[148] Readers may be interested to compare the similarities of this passage to *Georgics* 1.356–64 (which will be encountered again below), Cicero *On Divination* 1.8 (where he quotes from his own Latin translation of Aratus) and Pliny 18.85.359–60. There is also similarity to *On Weather Signs* 29, where we learn that a dog rolling is indicative of violent wind. And, in Aratus' *Phaenomena*, as in

On Weather Signs (28), birds can provide an indication of the weather (913–19): 'Also when a heron in irregular flight comes in from the sea to dry land uttering many a scream, it will be moving before a wind that is stirring over the sea. And sometimes too when petrels fly in fair weather, they move in flocks to face oncoming winds. Often wild ducks or sea-diving gulls beat with their wings on the land. . . .'[149] In some cases animal behaviours may be more useful as signs than astronomical observations: 'now when wasps in autumn swarms are massing everywhere, even before the setting of the Pleiades, one can tell the onset of winter, such is the whirling that suddenly eddies in the wasps'.[150] There were other noteworthy phenomena (*Phaenomena* 921–4): 'Before now the fluffy down of the white thistle's old age has been a sign of winds, when it floats abundantly on the surface of the mute sea, some of it drifting ahead, some behind.'[151] Virgil (*Georgics* 1.368–9) pointed to similar signs.[152]

Dalzell poses the question of the intended audience for Aratus' *Phaenomena* and makes some interesting remarks:

> No one is named in the poem, but an anonymous addressee is constantly being urged to take notice and pay attention. The demanding imperatives create a monotonous pattern in the poem, especially in its final section, but it is never made quite clear for whom all this instruction is intended. Farmers and sailors are the two groups who would benefit most from the information which the poem provides, and their needs are kept constantly before us. But the poem is not addressed to them. All we know about the addressee is that he might undertake a voyage and that he is expected to be familiar with the astronomical details of the Metonic cycle. Clearly it would be absurd to believe that the poem was really intended for the enlightenment of husbandmen and sailors. The large part which they play in the poem is more likely a conscious attempt to establish a link with Hesiod, who dealt both with farming and with seafaring.[153]

In Dalzell's view, Aratus did not intend his poem to be read by a specialist audience, with extensive knowledge or experience relevant to the subject. He notes that:

> Aratus' commitment to the intricacies of his subject has its limits. When it comes to discussing the working of the planets, he says, 'My courage fails' (460) and passes on to more

practical matters. The implied reader of the poem, therefore, is neither a scholar with a scientific interest in astronomy nor a farmer in need of advice about the weather. The poem pre-supposes one kind of reader and is addressed to another. This situation is typical of didactic poetry in general. The didactic poet speaks over the head of the formal addressee to a wider audience, whose identity has to be reconstructed from the text of the poem.[154]

Dalzell explains that 'a didactic poem always implies two kinds of reader: the immediate pupil to whom the poem is addressed and the true reader to whom this sophisticated kind of poetry will appeal'.[155]

Whatever the intended audience, the popularity of Aratus' poem made it a source of meteorological lore echoed by later writers. The *Phaenomena* was translated by various Roman writers into Latin. Terentius Varro Atacinus (born 82 BCE) produced a didactic poem using Aratus' work; the poem (which does not survive) was primarily concerned with weather-forecasting.[156] Cicero's Latin translation of the *Phaenomena* will be discussed in more detail below. A later writer, often identified as Germanicus Iulius Caesar, also produced a Latin version of Aratus' work.[157]

Germanicus was the elder son of Nero Claudius Drusus and Antonia, and was adopted (a common Roman practice) by his uncle, the emperor Tiberius. Tiberius himself had been adopted by Augustus; thus Germanicus was placed in the direct line of succession within the Julian lineage. In addition to his political and military career, Germanicus found time to pursue literary interests, and produced a number of works, including Greek comedies (now lost).

In the Latin poem attributed to Germanicus, there are many differences from that of Aratus (including 'corrections', apparently based on Hipparchus' commentary). Germanicus' work may be regarded, to some extent, as a paraphrase. In addition to lines 1–725, which comprise the bulk of the poem, five smaller fragments survive (in various manuscripts). While several of these are largely concerned with weather, the fragments are not based on the section of Aratus' poem which deals with weather signs. It is not clear whether Germanicus was relying on another source here, or whether the fragments on weather represent his own work and interests. As D.B. Gain, an editor of Germanicus' text, has explained, 'whether [the fragments concerned with weather] are based on another writer, or are a compilation of several sources, and whether there are original elements in them or not, is unknown'.[158] In fragment ii, Germanicus explains the path of the sun through the zodiac, and discusses the planetary motions. Fragments iii and iv contain details of the effects of each planet and sign on the weather. So, for example:

> The Ram scatters dreary rain mixed with hail and falling snow over the nearby ridges. The Bull carries water and arouses violent winds. Under him Jupiter often casts his thunderbolts, the sky is violent, fires are sent from it, and it thunders. Under the Twins winds gently caress the azure sky and moisture seldom travels down from sky to earth. Everything grows mild under the peaceful sign of the Crab. The Lion is dry, seeing that his breast is particularly hot.[159]

Because so much of this part of the poem is lost, it is difficult to say much about the author's interest in weather.

Nevertheless, the opening lines and dedication of the poem, while they echo and, indeed, mention Aratus directly, also point to some differences between the interests of the two poets. Germanicus' poem begins this way:

> Aratus began with mighty Jupiter. My poem, however, claims you, father, greatest of all, as its inspirer. It is you that I reverence; it is to you that I am offering sacred gifts, the first fruits of my literary efforts.

He continues:

> What power would there be in the points which mark for certain the seasons of the year, the one where the violent sun turns around in the sign of the burning Crab, the one where he grazes the opposite turning post in chill Capricorn, or those where the Ram and the Balance make the two divisions of the day equal, if the gaining of peace under your leadership had not allowed ships to sail the level sea, the farmer to till the land and the sound of arms to recede into distant silence? At last there is an opportunity to lift one's gaze boldly to the sky and learn of the celestial bodies and their different movements in the heavens and discover what the sailor and the canny ploughman should avoid, when the sailor should entrust his ship to the winds and the ploughman his seed to the soil.[160]

Toohey has argued that, as the heir to the Augustan succession and the Augustan peace, Germanicus, a military and political leader as well as a poet, had a quite different agenda from that of Aratus. While Aratus' poem honours the universal order and harmony imposed by Zeus on the physical world, Germanicus' poem celebrates the peace of Augustus, without which one would not know when it is appropriate to use sails rather than oars (154–6), when to avoid being on the sea at all (288–305), when crops fail and when they flourish (335–40), and how to know the length of day and night (573–4). On Toohey's reading, Germanicus' version is an imperial praise poem, celebrating Rome, rather than a hymn to Zeus.[161] Nevertheless, the practical motivation for astronomical knowledge and weather prediction is clearly present.

A much later Latin version of Aratus' poem is the work of Avienus (or Avienius, fl. mid-fourth century CE; his full name appears to have been Postumius Rufius Festus Avienus[162]). In addition to his work *Aratea* (a paraphrase of the *Phaenomena* of Aratus),[163] he also translated the geographical poem of Dionysius Periegetes (second century CE) in his *Descriptio Orbis Terrae* and composed another didactic poem *Ora Maritima*. Avienus' version is dedicated to Jupiter. Hubert Zehnacker has argued that the poem should be considered, along with Avienus' other didactic poems, as part of his effort to describe the entire universe, heaven, earth and sea.[164]

Of the three surviving Latin versions of Aratus' *Phaenomena*, that of Avienus is the most complete; furthermore, it is the only Latin version in which the entire section on weather prognostication is extant. But it should be emphasized that Avienus' version is not merely a translation. As Jean Soubiran and, following him, Zehnacker, have both noted, Avienus' poem is much longer than that of Aratus; this is credited, in part, to Avienus' verbosity.[165]

Aratus' poem has 1,154 lines, that of Avienus 1,878. A comparison of parallel sections of the two poems is interesting. Zehnacker has calculated that for every line in Aratus' version which describes the constellations (there are 432 such lines in Aratus), Avienus wrote nearly twice as many (831 lines); for each of Aratus' lines of poetry on the planets, the celestial circles, and risings and settings, Avienus provides 1.5 lines. However, in the section on meteorological prognostication, Avienus is less forthcoming, writing only 1.3 lines for each of Aratus. Both Soubiran and Zehnacker comment on what appears to be Avienus' waning interest as he nears the end of his work. Soubiran argues that Avienus seems to have been attempting to produce a more 'scientific' work than that of Aratus, following the model of Lucretius (whose poem will be considered in the next chapter, which deals with explanation of meteorological phenomena). To this end, Avienus did not hesitate to cut material from Aratus' section on weather signs, and to add material from the scholiasts. Soubiran outlines Avienus' insertions into the Aratean framework, noting that his interests in astronomy and physics dominate, while he downplays (and cuts) references to animal behaviour.[166] Further, Avienus does not restrict himself to listing weather signs; in places, he offers an explanation of various meteorological and astronomical phenomena. So, for example, in lines 1,409–45, Avienus discusses the origin of clouds, of winds and of rain; in lines 1,631ff. he discusses the cause of solar eclipses. These explanatory insertions are a notable difference between Aratus' poem and Avienus' Latin version.

Aratus' *Phaenomena* was certainly not the first to give information about constellations and weather signs. As was noted earlier, the final portion of the Hesiodic *Works and Days* is a sort of farmers' almanac in verse form, providing information about astronomical phenomena, such as the appearances of particular constellations, and associated seasonal weather conditions. The Roman poet Virgil composed a poem in four books, centred on agricultural themes, hence known as the *Georgics*. The poet is not necessarily concerned with the professional training of farmers, but with the broader issues of the attainment of happiness.[167] Aratus' poem was an obvious source for the information on weather contained in the first book of the *Georgics*, but the *Works and Days* can also be seen as a model for Virgil on several levels.[168] The work known as *On Weather Signs* may also have been utilized by Virgil.[169]

In the *Georgics* (1.316ff.), Virgil describes outbursts of violent weather:

> Well, often I've seen a farmer lead into his golden fields
> The reapers and begin to cut the frail-stalked barley,
> And winds arise that moment, starting a free-for-all,
> Tearing up by the roots whole swathes of heavy corn
> And hurling them high in the air: with gusts black as a
>   hurricane
> The storm sent flimsy blades and stubble flying before it.
> Often, too, huge columns of water come in the sky
> And clouds charged off the deep amass for dirty weather
> With rain-squalls black: then the whole sky gives way,
>   falls,
> Floods with terrific rain the fertile crops and the labours

Of oxen; filled are the ditches, dry rivers arise in spate
Roaring, the sea foams and seethes up the hissing fjords.

Such weather wreaks agricultural havoc, and Virgil warns readers
(335), 'in fear of this, mark the months and signs of heaven'. He
explains that (351ff.), 'through unfailing signs we might learn these
dangers – the heat, and the rain, and the cold-bringing winds – the
father himself decreed what warning the monthly moon should give,
what should signal the fall of the wind, and what sight, oft seen,
should prompt the farmer to keep his cattle nearer to their stalls'.[170]
Virgil (1.356–9) goes on to provide details of some of these signs:

At once, when winds are rising,
The sea begins to fret and heave, and a harsh crackling
Is heard from timbered heights, or a noise that carries far
Comes confused from the beaches, and copses moan
    crescendo.[171]

(Readers will recognize similar passages in other authors mentioned
above, for example *On Weather Signs* 29 and 31, Aratus *Phaenomena*
909, Cicero *On Divination* 1.8 and Pliny *Natural History* 18.359–60.)
Virgil's weather signs seem to be very traditional, incorporating
those listed by other authors. But Virgil is not merely providing a
list of weather signs; his description is memorable for its vividness
and for the variety of signs of weather described in close juxtaposi-
tion. He provides a lively and suggestive evocation of weather
prognostication. So, for example, in the lines which follow on from
the last passage (1.360ff.):

At such a time are the waves in no temper to bear your
    curved ship –
A time when gulls are blown back off the deepsea flying
Swift and screeching inland, a time when cormorants
Play on dry land, and the heron
Leaves his haunt in the fens to flap high over the cloud.
Another gale-warning often is given by shooting stars
That streak downsky and blaze a trail through the night's
    blackness [cf. Aratus 926]
Leaving a long white wake:
Often a light chaff and fallen leaves eddy in the air,
Or feathers play tig skimming along the skin of water.
    [Cf. *On Weather Signs* 37]

But when lightning appears from the quarter of the grim
    north wind,
When it thunders to south or west, then all the countryside
Is a-swim with flooded dykes and all the sailors at sea
Close-reef their dripping sails.[172] [Cf. Aratus *Phaenomena*
    933; *On Weather Signs* 21]

Virgil (1.373ff.) explains that 'rain need never take us unawares';
animals provide many signs of impending weather:

for high-flying cranes will have flown to valley bottoms
To escape the rain as it rises, or else a calf has looked up
At the sky and snuffed the wind with nostrils
    apprehensive,
Or the tittering swallow has flitted around and around the
    lake,
And frogs in the mud have croaked away at their old
    complaint. [Cf. *On Weather Signs* 42]
Often too from her underground working the emmet,
    wearing
A narrow path, bears out her eggs [cf. Aratus 956]; a giant
    rainbow
Bends down to drink; rook armies desert their feeding-
    ground
In a long column, wing-tip to wing-tip; their wings
    whirring.
Now seabirds after their kind, and birds that about
    Caÿster's
Asian waterflats grub in the fresh pools, zestfully fling
Showers of spray over their shoulders,
Now ducking their head in the creeks, scampering now at
    the wavelets,
Making a bustle and frivolous pantomime of washing.
Then the truculent raven full-throated announces rain
As she stalks along on the dry sand.

But not all of the non-astronomical signs are related to animal
behaviour; for 'even at night, maidens that spin their tasks have not
failed to mark a storm as they saw the oil sputter in the blazing lamp,
and a mouldy fungus gather on the wick' (*Georgics* 390; cf. Aratus
977f.).[173]

Virgil concludes the section on weather signs (1.460ff.) by noting the more general (and political) significance of weather, its effects and assumed portents:

> Lastly, what the late evening conveys, from whence the
>   wind drives
> Fine-weather clouds, and what the damp south wind is
>   brooding,
> The sun discloses. Who dares call the sun a liar?
> Often too he warns you of lurking imminent violence,
> Of treachery, and wars that grow in the dark like a cancer.
> The sun, when Caesar fell, had sympathy for Rome –
> That day he hid the brightness of his head in a rusty fog
> And an evil age was afraid his night would last for ever.[174]

It is tempting to claim that the poet is overusing his licence here, but at least one translator of the poem, H.R. Fairclough, reports that 'historians, as well as poets, assure us that the atmospheric conditions of the year 44 BCE (the year of Caesar's assassination) were remarkable'.[175]

Dalzell points out that modern critics have often argued that didactic poems like Virgil's *Georgics* and Aratus' *Phaenomena* are actually of minimal practical value, having little to do, actually, with husbandry and sea-faring. Some critics have argued that, insofar as this is the case, such poems must not be seriously regarded as didactic. But as Dalzell rightly notes:

> It does not follow, however, that because the poem is not a practical manual, it is not in some sense a didactic work. There is, after all, more than one kind of didacticism. One should not assume that, because a work is not designed for the professional, it lacks didactic interest. Our bookshops are full of volumes which explain to a general audience how various tasks are to be performed. I read recently in an English newspaper that books on cooking regularly outsell such popular genres as mystery and romance. Statistics like this should alert us to the fact that didactic works have always had a wide appeal. As we have already seen, Aratus' poem on astronomy enjoyed an extraordinary success and inspired at least twenty-seven commentaries and four translations into Latin. Agriculture was a subject with an even wider appeal in antiquity. Varro was able to cite more than fifty Greek

works which dealt with various agricultural matters (*de re rustica* 1.1.7–8). Pliny tells us that after the fall of Carthage the city's libraries were scattered among the local rulers, but the twenty-eight volumes of Mago's great work on agriculture were preserved and translated into Latin by order of the Senate (18.22).[176]

The thought of 'didactic' works may at first sound rather dull, but as Dalzell indicates, such volumes have broad appeal. While it would be anachronistic to attempt to draw too many parallels between the audiences for ancient and modern didactic works, the popularity of the genre was well attested in antiquity, as it continues to be today.

Peter Toohey has argued that Hesiod's *Works and Days* is the prototypical didactic poem; many of its stylistic features were adopted and incorporated by later didactic poets.[177] Didactic poems tend to share certain elements, including specific forms of address and argumentation; quasi-formulaic expressions are also employed.[178]

It is worth considering the use of formulaic expressions in the didactic poems that teach about weather signs, including Aratus' *Phaenomena* and Virgil's *Georgics*. Certainly, it may be the case that the information concerning weather signs is very similar simply because writers are describing what were regarded as recognized indications of weather. Further, it is very likely that later authors were, to some extent, using the work of their predecessors as a source of information. In contrast to the *parapēgmata*, earlier authorities are not cited by name; this may be a poetic convention for, in poetry, 'authorities' are rarely cited,[179] although there may be references to predecessors. We find frequent echoes of earlier writers, even when their names are not mentioned.

It is possible that later writers were echoing their predecessors, not only in terms of information, but also in form (and even formulae). There was a great deal of 'intertextuality' at work between the authors of the various texts (both prose and poetic). Of course, the value of teaching formulaically expressed weather signs may be variously assessed. There might be a modern tendency to 'pooh-pooh' formulaic expressions as 'old-wives' tales', but within Hellenistic culture such poetry was highly valued. Further, it is not at all clear that such formulae would have been discounted; rather, echoes and resonances with earlier authors may have been recognized as expressions of 'received' wisdom. What is more, many of the authors specifically refer to their predecessors and their texts as authorities. Pliny's reliance on the authority of Cato, Virgil and Caesar is an example.

But though there is abundant evidence of a tradition, both in poetry and prose, of an interest in weather signs, not everyone accepted signs of impending weather unquestioningly. Cicero translated Aratus' *Phaenomena* into Latin. He quotes passages from his translation of the section on weather signs and situates them within a more general discussion of divination of various sorts, in a dialogue *On Divination*.[180]

In this work, Cicero and his brother Quintus consider the various forms of divination practised by Greeks, Romans and Chaldeans. The work opens with Cicero explaining that 'there is an ancient belief, handed down to us even from mythical times and firmly established by the general agreement of the Roman people and of all nations, that divination of some kind exists among men; this the Greeks call μαντική [mantikē] – that is, the foresight and knowledge of future events'.[181] Cicero then presents a brief overview of the history and various types of divination, noting that philosophers had gathered 'certain very subtle arguments to prove the trustworthiness of divination' (1.3).[182] He admits that he was 'in doubt as to the proper judgement to be rendered in regard to divination', whether to accept or reject the claims that divination can accurately predict future events. He hesitates, acknowledging that 'we run the risk of committing a crime against the gods if we disregard them, or of becoming involved in old women's superstition if we approve them' (1.4).[183]

In the dialogue that follows, Cicero and his brother consider various divinatory practices and claims. Quintus discusses the use of weather signs at some length, even though he makes it clear that this is not, strictly speaking, a form of divination. Here (1.7) he quotes a passage, Cicero's own Latin translation of Aratus:

The heaving sea oft warns of coming storms,
When suddenly its depths begin to swell;
And hoary rocks, o'erspread with snowy brine,
To the sea, in boding tones, attempt reply;
Or when from lofty mountain-peak upsprings
A shrilly whistling wind, which stronger grows
With each repulse by hedge of circling cliffs.[184]

Quintus (1.8) asks his brother Cicero, 'Your book, *Prognostics* [*Prognostica*], is full of such warning signs, but who can fathom their causes?'[185] He explains that a Stoic, Boëthus, attempted to do so and did succeed in explaining phenomena related to the sea and the sky.

But Quintus then gives a list (once again in a Latin translation from Aratus) of signs of weather provided by the behaviour of birds and asks: 'who can give a satisfactory reason why the following things occur?' He notes that 'Hardly ever do we see such signs deceive us and yet we do not see why it is so.' Pointing to the ability of frogs to indicate the coming weather, he enquires: 'who could suppose that frogs had this foresight? And yet they do have by nature some faculty of premonition, clear enough of itself, but too dark for human comprehension.' Quintus concludes by affirming that 'I do not ask why, since I know what happens.'[186] He goes on to discuss the practices which he does consider to be divination, and concludes by noting that 'believing as I do that the gods do care for man, and that they advise and often forewarn him, I approve of divination which is not trivial and is free from falsehood and trickery'.[187]

In the second part of the dialogue (Book 2), Cicero answers his brother, refuting the idea that divination is of any value. It is clear from remarks made in the course of his argument that he considers trained specialists to have knowledge and skills not held by diviners. Furthermore, in some cases these specialists do engage in prediction of a particular type, for example physicians predict the course of disease and sailors (pilot = *gubernator*) must judge weather conditions. Cicero asks (2.5): 'do you think that a prophet will "conjecture" better whether a storm is at hand than a pilot? or that he will by "conjecture" make a more accurate diagnosis than a physician, or conduct a war with more skill than a general?'[188] He addresses (2.5) this distinction between the skill of trained specialists and the conjectures of diviners, by arguing against Quintus' earlier remarks about weather prediction:

> You went on to say that even the foreknowledge of impending storms and rains by means of certain signs was not divination, and, in that connexion, you quoted a number of verses from my translation of Aratus. Yet such coincidences 'happen by chance' for though they happen frequently they do not happen always. What, then, is this thing you call divination – this 'foreknowledge of things that happen by chance' – and where is it employed? You think that 'whatever can be foreknown by means of science, reason, experience, or conjecture is to be referred, not to diviners, but to experts.' It follows, therefore, that divination of 'things that happen by chance' is possible only of things which cannot be foreseen by means of skill or wisdom.[189]

Cicero goes on (2.6) to ask 'Can there, then, be any foreknowledge of things for whose happening no reason exists?' He responds to his own question, explaining that:

> we do not apply the words 'chance,' luck,' 'accident,' or 'casualty' except to an event which has so occurred or happened that it either might not have occurred at all, or might have occurred in any other way. How, then, is it possible to foresee and to predict an event that happens at random, as the result of blind accident, or of unstable chance? By the use of reason the physician foresees the progress of a disease, the general anticipates the enemy's plans and the pilot forecasts the approach of bad weather. And yet even those who base their conclusions on accurate reasoning are often mistaken.

He concludes this section by arguing that 'if then mistakes are made by those who make no forecasts not based upon some reasonable and probable conjecture, what must we think of the conjectures of men who foretell the future by means of entrails, birds, portents, oracles, or dreams?'[190]

This discussion between Cicero and Quintus regarding the value of divination is important, because it indicates that weather prognostication using astronomical methods was not necessarily regarded in a similar way to divinatory practices such as dream interpretation. Remarks attributed to Quintus make it clear that even those who valued divination did not consider astrometeorology to be on a par with divination; those who argued against the value of divination emphasized the reliance on reason and experience used by sailors in weather prediction. But it is interesting that Quintus felt the need to explain that astrometeorology was not divinatory; his disclaimer suggests that some people may have regarded it as such.

Vitruvius (first century BCE), in his discussion of *astrologia* (Book 9, *On Architecture*) provided a brief list of ancient Greeks who offered theories explaining nature and those who used astronomical *parapēgmata*. He notes that:

> In natural philosophy, Thales of Miletus, Anaxagoras of Clazomenae, Pythagoras of Samos, Xenophanes of Colophon, Democritus of Abdera left elaborate theories on the causes by which nature was governed, and the manner in which each produced its effects. Eudoxus, Eudemus, Callippus, Meton,

Philippus, Hipparchus, Aratus, and others followed up their discoveries, and, with the help of astronomical tables [*ex astrologia parapegmatorum*], discovered the indications of the constellations, of their setting, and of the seasons, and handed down the explanations to after times. Their knowledge is to be highly regarded by mankind, because they so applied themselves, that they seem by divine inspiration to declare beforehand the indications of the seasons.[191]

Thus, some parapegmatists may have seemed to have been so clever at weather prediction that they were regarded by others as having divinatory skills.

Astrometeorology, as a type of astrology, was apparently regarded, at least by some Romans, as a type of divination.[192] While Roman interest in astrology is a subject much broader than our concerns here, it is worth returning to Cicero, and his discussion of astrology, insofar as it relates to weather prediction. Cicero discusses astrology at several places in *On Divination*. Astrology is the first type of divinatory practice mentioned. At the beginning of the work (1.1), he comments that 'I am aware of no people, however refined and learned or however savage and ignorant, which does not think that signs are given of future events, and that certain persons can recognize those signs and foretell events before they occur.' He went on to explain that:

the Assyrians, on account of the vast plains inhabited by them, and because of the open and unobstructed view of the heavens presented to them on every side, took observations of the paths and movements of the stars, and, having made note of them, transmitted to posterity what significance they had for each person. And in that same nation the Chaldeans – a name which they derived not from their art but their race – have, it is thought, by means of long-continued observation of the constellations, perfected a science which enables them to foretell what any man's lot will be and for what fate he was born. The same art is believed to have been acquired also by the Egyptians through a remote past extending over almost countless ages.[193]

Later, in Book 2, Cicero argues quite strenuously against astrology, the 'Chaldean method' of foretelling the future. He explains (2.42) that:

since, through the procession and retrogression of the stars, the great variety and change of the seasons and of temperature take place, and since the power of the sun produces such results as are before our eyes, they believe that it is not merely probable, but certain, that just as the temperature of the air is regulated by this celestial force, so also children at their birth are influenced in soul and body and by this force their minds, manners, disposition, physical condition, career in life and destinies are determined.[194]

But he complains (2.43) 'What inconceivable madness! For it is not enough to call an opinion "foolishness" when it is utterly devoid of reason.'[195] He later remarks (2.45), 'what utter madness in these astrologers in considering the effect of the vast movements and changes in the heavens, to assume that wind and rain and weather anywhere have no effect at birth!' He asks the rhetorical question: 'in view of the fact that the heavens are now serene and now disturbed by storms, is it the part of a reasonable man to say that this fact has no natal influence – and of course it has not – and then assert that a natal influence is exerted by some subtle, imperceptible, well-nigh inconceivable force which is due to the condition of the sky, which condition, in turn, is due to the action of the moon and stars?'[196] But although Cicero discredits natal astrology, and predictions as they apply to individuals, he apparently does not dismiss the physical effects of the celestial bodies on the atmosphere. For Cicero, weather prediction was based on astronomical knowledge, was not divination and so was acceptable (if not always accurate), whereas astrological prediction of human events was divinatory and completely disreputable.

Other authors shared this view, that prophecy about the fate of individuals was not possible through astronomical divination, while at the same time utilizing astronomical methods and tools, including *parapēgmata*, for predicting weather phenomena and associated effects. For example, in his *Natural History* (2.6.29), Pliny asserted that 'there is no such close alliance between us and the sky that the radiance of the stars there also shares our fate of mortality'.[197] Nevertheless, he provided detailed information correlating astronomical phenomena to weather. And he did not discount the effects of celestial bodies on terrestrial life. He notes (*NH* 2.41.108) that 'the parts of some constellations have an influence of their own – for instance at the autumnal equinox and at mid-winter, when we learn by the storms that the sun is completing its orbit; and not only by falls of rain and storms, but by many things that happen to our bodies and to the fields'. For example:

> Some men are paralysed by a star, others suffer periodic disturbances of the stomach or sinews or head or mind. The olive and white poplar and willow turn round their leaves at the solstice. Fleabane hung up in the house to dry flowers exactly on midwinter day, and inflated skins burst. . . . Indeed persistent research has discovered that the influence of the moon causes the shells of oysters, cockles and all shellfish to grow larger and again smaller in bulk, and moreover that the phases of the moon affect the tissues of the shrewmouse, and that the smallest animal, the ant, is sensitive to the influence of the planet and at the time of the new moon is always slack.[198]

While Pliny is not always the most consistent author (he relied on the works of a large number of other writers in compiling his *Natural History*), his denial of the ability to predict the fate of individuals from stellar observations does not argue against his assertion of the influence of celestial events on earthly life.[199]

Pliny was no poet, yet it may be argued that his *Natural History* is a didactic work that has close resonance with some of the previously mentioned poems. For example, many of the same weather signs are mentioned and discussed by Pliny as by earlier authors; he utilized their work and, in some cases, referred to his predecessors by name. And the valuing of the virtues of hard work and the agricultural life are themes shared with the *Works and Days* and the *Georgics*.

It is worth considering what role prediction plays in these various works. Roger French, in his book *Ancient Natural History*, has made the point that meteorological phenomena were very familiar, and potentially useful or frightening to the ancients (as they still are to us today). Traditionally, meteorological phenomena were linked to the gods, often as their epiphanies. Zeus made rain, Poseidon caused the earth to tremble in quakes, Iris appeared as a rainbow. But clearly, the relatively early evidence of the *Works and Days*, as well as the other texts cited here, shows a desire to predict and utilize meteorological phenomena, particularly weather, without reference to gods. French has argued that 'the denial of the gods marked out the shape of natural philosophy: their absence defined what "nature" was'. He notes that 'the many different explanations by the philosophers of lightning had in common the principal feature that they did *not* attribute it to Zeus, as non-philosophical Greeks thought'. When meteorological phenomena are not regarded as linked to the gods, they can be explained as natural phenomena, with reference only to other natural phenomena, completely side-lining the gods. French suggests that, as a subject of study, natural philosophy 'has not only had its boundaries fixed but its subject-matter determined by the absence of the actions of the gods in things that frightened people'.[200]

Taking this line further, it may be argued that the prediction of meteorological phenomena with reference to other natural phenomena, including the activities and motions of the sun, moon, stars, birds, frogs and cattle, is part of a conception of the 'natural' world. The various weather signs are 'natural' and not attributed to the gods. Observation of these 'natural' weather signs is possible; detailed lists and instructions can be provided and followed. These details add an important measure of perceived control, even when it is acknowledged that it may not be complete.

Other ancient authors also provide information about weather signs and their uses. For example, Claudius Aelianus (known as Aelian, 165/70–230/5 CE) recorded extraordinary stories and anecdotes about animals in his work *On the Characteristics of Animals*; some of these stories include the abilities of various animals, including birds, to predict the weather. Aelian names some of his sources, including Aristotle.[201] Aelian (7.7) credits Aristotle with communicating the information that 'cranes flying in to land from the sea indicate to the intelligent man that a violent storm is threatening' and notes (7.8) that 'the Libyans are equally bold in stoutly maintaining that in their country the goats . . . also give clear signs of impending rain'.[202] His examples of weather prediction are not restricted to animals.

Famous mathematicians and philosophers are also credited with having been able to predict the weather accurately, though this skill may have been linked to wearing the right sort of garment. Aelian relates the following stories of humans predicting the weather successfully: 'Hipparchus in the reign of Hiero the Tyrant was sitting in the theatre wearing a leathern jerkin, and astonished people by knowing in advance out of the clear weather then prevalent that a storm was coming. And Hiero in his admiration of the man congratulated the people of Nicaea in Bithynia on having Hipparchus as a citizen.' Similarly, 'when at Olympia Anaxagoras, likewise clad in a leathern jerkin, was watching the Olympic Games and a storm of rain burst, all Hellas sang his praises, and claimed that his wisdom was more that of a god than of a man'. These heroic anecdotes are followed by examples of animal behaviour useful for predicting weather, including the information that 'if an ox bellows and sniffs the air, rain is inevitable' and 'when pigs appear in cornland, they inform us that the rain is departing'. Strikingly, Aelian concludes this section on weather signs with the following remark: 'In all these matters men fall behind: they only know these changes when they occur.'[203] Aelian seems to be suggesting that such animal behaviours are not only useful for humans; they are understood by the animals exhibiting the behaviour.

Weather phenomena were sometimes used as signs to predict other events; they could themselves be used in divination. The practices of divination rely on the belief that the gods send messages to humans, by means of signs. Various types of signs were recognized as divinatory and various means were practised for their recognition and interpretation (for some required interpretation). As Jerzy Linderski has explained, the Roman states employed three different types of experts in divination: the board of priests who were in charge of the Sibylline oracular books, the *augures* (augurs), and the haruspices, who specialized in examining the entrails of sacrificial victims. But both the augurs and the haruspices observed and interpreted signs based on the behaviour of birds and of celestial events, including *fulmina* (lightning) and *tonitrua* (thunder).[204] The augurs and haruspices interpreted these signs differently, but both groups utilized meteorological phenomena as indications of divine will.

Cicero has the Stoic speaker Balbus, in his dialogue *On the Nature of the Gods* (2.5), address the role of divine signs within Roman life:

> lightning, storms, rain, snow, hail, floods, pestilences, earthquakes and occasionally subterranean rumblings, showers of stones and raindrops the colour of blood, also landslips and chasms suddenly opening in the ground . . . the appearance of meteoric lights and what are called by the Greeks 'comets', and in our language 'long-haired stars', such as recently during the Octavian War appeared as harbingers of dire disasters, and the doubling of the sun, which my father told me had happened in the consulship of Tuditanus and Aquilius, the year in which the light was quenched of Publius Africanus, that second sun of Rome: all of which alarming portents have suggested to mankind the idea of the existence of some celestial and divine power.[205]

French has commented on this passage, arguing that meteorological phenomena were discussed because they were so interesting, important and frightening; and because they were commonly regarded as being produced by the gods, as well as providing evidence for the warnings of chastisement.[206] He notes that the Roman Stoic Balbus restores, rather than removes, fear of the gods. Cicero represented several different conflicting points of view in the dialogue; as we have seen, some people, including Cicero, rejected the idea that the gods send portents. Nevertheless, it is interesting that some types of signs (whether divinely communicated or not) were indicative of future weather, while weather phenomena themselves also served as signs of other future events. Meteorological phenomena were of vital concern, being continually present, often useful and sometimes frightening. The discussions presented in Cicero's dialogues illustrate the lack of universal agreement about the best way to regard meteorological phenomena, and indicate a degree of ambivalence concerning the role of the gods.

The prediction of meteorological phenomena through various types of observations and signs is well attested in a wide range of sources. In addition to the *parapēgmata*, didactic poems, encyclopedic histories and astronomical works considered above, information regarding weather signs survives in other kinds of texts. For example, Book 26 of the pseudo-Aristotelian *Problems* is concerned with topics connected with the winds. The formulation of some of the questions indicates that certain phenomena were regarded as weather signs. So, for example, question 23 is phrased: 'Why, when there are shooting stars, is it a sign of wind?'[207] Similarly, Seneca's *Natural Questions* is concerned with meteorological phenomena. While these texts do give some information regarding what phenomena were regarded as weather signs, their authors seem more interested in explanation than in prediction.[208]

A late author, John Lydus (John the Lydian, 490–c. 560 CE) wrote a work devoted to signs. Lydus was a civil servant at Constantinople. Later in life (probably following his retirement in 551/2),[209] he wrote three works, *De Magistratibus*, *De Mensibus* and *De Ostentis*, which largely survive; his more youthful writings do not. While all three works show his somewhat antiquarian interests, they also indicate his deep involvement with contemporary affairs. *De Magistratibus* is concerned with the history and workings of the administrative offices of the Roman state, meant to serve as a guide for the restoration of its former glory.[210] The other two works are compilations of material by earlier authors; both have an antiquarian flavour. *De Mensibus* deals with calendars and festivals;[211] *De Ostentis* is concerned with signs and portents.

The contents of *De Ostentis* may be summarized as follows: Introduction (1–8); On events signified by the sun and the moon (9a–d); On comets (10a–10b); Dissertation of Campester on comets (11–15); General observations pertaining to the moon, distinguished according to the lunar months (17–20); On thunder (21–22); Treatise on thunder from the teachings of the Egyptians, distinguished according to solar months (23–6); Daily divination from thunder, regional with reference to the moon, according to the Roman Nigidius Figulus from the writings of Tages by way of verbatim translation (27–38); Divination from thunder from the writings of Fonteius the Roman by way of verbatim translation (39–41); General observation with reference to the moon from the summer solstice from the writings of Labeo by way of verbatim translation (42); On thunderbolts (43–6); Labeo's *Liber Fulguralis* (47–52); On earthquakes (53–4); Seismology of Vicellius (55–8); Daily record of the entire year from the writings of Claudius Tuscus by way of verbatim translation (59–71H); and Astrological ethnology (71).[212] While some scholars, notably Anastasius C. Bandy and Michael Maas, have emphasized his interest in astrology,[213] Lydus' compilation can also be understood as a collection of meteorological signs; the inclusion of signs specifically related to earthquakes and seismological activity is particularly interesting. He provides information culled from authors whose works are otherwise lost.

Maas emphasizes Lydus' work as being self-consciously part of a tradition; he explains that:

> Two general assumptions underlie *de Ostentis*: that the learning of the past is still valid and that general celestial principles can be seen to operate through their earthly manifestations. Lydus posits no discontinuity with the scientific tradition of which he understood himself to be a part. When he adds his own observations and conclusions to the extensive excerpts he has culled from the astrological corpus, he affirms their validity, continued utility, and reliability. He is, in effect, deliberately contributing to a long scientific tradition. Lydus does not hope to add any new twists, only to demonstrate that the old principles still hold true.[214]

Lydus was avowedly only interested in investigating the possibility of prediction; he aimed to 'see if anything is to be gained from these signs to predict the possible outcome of future events'.[215] To some extent, he positioned himself as a defender of the tradition of astro-meteorological prediction; he claimed that his examples were 'in reply to those who object to signs and dare to speak against Ptolemy'.[216] Interestingly, he saw a strict division of labour between prediction and explanation. He took pains to emphasize that 'we are not concerned to explain the physical causes of these phenomena or to discuss theories about them – that is for the philosophers to worry about'.[217] And, as we will see in the following chapters, philosophers (as well as other authors) did indeed concern themselves with the causes of meteorological phenomena.

3

EXPLAINING DIFFICULT
PHENOMENA

As we have seen, some ancient authors believed that there should be a strict division between those who sought to explain and discuss theories of meteorological phenomena, and those who aimed to make predictions. The last chapter focused on ancient approaches to prediction. The texts examined there were concerned, to a large extent, with answering questions such as 'what will the weather be?' and 'what are the various signs and indications of particular types of weather?' For the most part, these texts do not discuss the causes of weather phenomena. But it is striking that both those interested in prediction and those committed to explanation strongly convey a sense of the difficulty and lack of certainty involved; this sense of coping with intractability and uncertainty characterizes ancient Greek and Roman approaches to meteorology. Weather prediction was not regarded as an infallible enterprise, and the inherent challenges of determining the causes of meteorological phenomena were emphasized. Nevertheless, ancient authors deliberately confronted the obstacles and offered explanations of the causes of meteorological events. In the texts considered next, the authors address questions such as 'how do different meteorological phenomena occur?' and 'what causes different sorts of weather?' By and large, the texts that discuss causes are not particularly concerned with prediction.[1]

There are very few surviving treatises devoted to the explanation of meteorological phenomena: most notably Aristotle's *Meteorology* and, arguably, Seneca's *Natural Questions*. Other works which focused on this subject, but are no longer fully extant, include Theophrastus' *Meteorology* (known also as *Metarsiology*, partially preserved in Arabic and Syriac). In their writings on meteorological topics, these three authors, Aristotle (384–322 BCE), Theophrastus (372/1 or 371/70–288/7 or 287/6 BCE) and Seneca (born 4 BCE–1 CE, died 65 CE), provide information about the ideas of others, as well as presenting their own theories. This provision of information about predecessors is an important feature of the tactics employed in ancient natural philosophy; the approach reinforces the notion that the work of each author is part of a 'joint' effort.

In fact, much of what is known about the meteorological ideas of many ancient philosophers survives only through descriptions, quotations and discussions presented in the works of successors. Each of these would have had a particular purpose in composing their own works and in making reference to earlier thinkers. These authors were not necessarily concerned with accurately preserving the ideas of their predecessors, although in some cases that was the intention. In other cases, the discussion of others' ideas provided a basis for developing the author's own point of view. Even when an author did aim to report accurately a predecessor's views, it is possible that his reading or interpretation may have altered the ideas of the earlier thinker to some extent.

Modern scholars must rely, in some cases almost completely, on reports of ancient thinkers' ideas transmitted in various ways (including paraphrase, as well as quotation) by other writers. These others were, in some instances, contemporaries, but some lived centuries later. The authors who reported the ideas of others did not always use, or have access to, the original writings and statements. Even in antiquity, authors relied on secondary sources, including summaries of the works of earlier thinkers, in handbooks and epitomes produced by others. Furthermore, the collection of the opinions (*doxai*) of philosophers, presented as a 'doxography', appeared in various formats; doxographies served to preserve ideas which otherwise might have been lost. How accurately authors quoted the views of others is difficult to judge; it is clear that those who presented opinions and views of other thinkers had their own motives for doing so. For our purposes, it is fortunate that the authors of the few surviving ancient treatises on meteorology often provide information about the ideas of earlier thinkers.[2]

The Presocratic philosophers

The ancient authors whose works on meteorology survive, namely, Aristotle, Theophrastus and Seneca, serve as our principal sources of information about the meteorological ideas of the Presocratics.[3] While it is difficult to say much about Presocratic views on meteorology with any confidence, it is worth considering some of those fragments and testimonies which do survive, for they offer a sense of the ideas attributed by later ancient authors to their predecessors. However, these reports may say more about the writers who provided the testimonies and fragments than about the thinkers whose ideas they were said to represent.

Aristotle's *Meteorology* survives as one of the most important ancient works on the topic and is the earliest surviving in its entirety. In addition to developing his own views, Aristotle provides information on the ideas of his predecessors. At different points throughout the work, Aristotle deliberately relates his ideas to those of others; it was an essential part of his method to survey the opinions of others. In some cases he is anxious to argue against the ideas of his predecessors; in others he is happy to show that he agrees. In another work, the *Metaphysics* (1.3, 983b6–27), Aristotle describes Thales of Miletus (fl. 586 BCE) as the founder of a group of early philosophers that theorized about nature. While Thales of Miletus is conventionally credited with many interests, including astronomy, his views on meteorological phenomena are not well attested.

The Roman Stoic Lucius Annaeus Seneca was a prolific writer, whose writings covered a wide range of topics. Towards the end of his life he wrote the *Natural Questions*, in which he provided a detailed discussion of many meteorological phenomena. In it Seneca gives a good deal of information about the views of his predecessors. He reports that Thales stated that 'the world is held up by water and rides like a ship, and when it is said to "quake" it is actually rocking because of the water's movement'.[4] Seneca makes it clear that he rejects Thales' reported ideas, calling them absurd or silly, and suggesting that they be rejected as 'antiquated and ignorant'.[5] It is impossible to be sure that Thales put forward such views on earthquakes, but there seems no immediate reason to doubt it. It is interesting that Seneca claims to be able to present an earlier explanation of earthquakes, which he dismisses as 'old-fashioned'. Seneca is interested in demonstrating his own knowledge of the long history of explaining such phenomena. Thales may be used here simply to represent outdated opinion and to serve as a straw man as Seneca develops his own discussion.

Seneca's dismissal of Thales' ideas ignores what is often claimed to have been the innovation of the Milesian philosophers, that is, the rejection of the gods as causing natural phenomena. Seneca may have taken for granted the attempt to ground explanations as natural, rather than supernatural. The idea that the earth floats on water is present in Egyptian and Babylonian myths;[6] traditional Greek myth credited the god of the sea, Poseidon, with the power to cause earth-quakes. Poseidon is strikingly absent from the explanation Seneca credits to Thales. The absence of the traditional gods is convention-ally taken as a hallmark of ancient Greek natural philosophy, yet Aristotle reports (*On the Soul* A5, 411a8), that Thales thought that 'all things are full of gods'. As we will see in the next chapter, many natural philosophers retained the idea of divinity within the cosmos, while offering rational, 'natural' explanations of meteorological phenomena, in which the gods played no direct role.

Several ancient authors offer accounts of the meteorological views of Anaximander of Miletus (fl. 547/6 BCE), who was said to have been a relation, student and successor of Thales.[7] The fourth-century BCE philosopher Theophrastus, who was himself interested in meteor-ology, wrote a work known as *Opinions of the Physicists,* or *Physical Opinions*. This is no longer extant, but portions survive as brief quotations and summaries in the work of other writers. In present-ing the ideas of others, it is possible that Theophrastus did not always resist the temptation to provide what he regarded as 'appropriate' explanations of particular natural phenomena, even when these explanations were not explicitly present in his sources.[8]

Aëtius (probably late first century CE) produced a survey of Greek natural philosophy, very likely indirectly derived from Theophrastus, but shaped and supplemented by others.[9] Under the heading 'On thunder, lightning, thunderbolts, whirlwinds and typhoons', he reports that 'Anaximander says that all these things occur as a result of *pneuma* [ἐκ τοῦ πνεύματος]: for whenever it is shut up in a thick cloud and then bursts out forcibly, through its fineness and lightness, then the bursting makes the noise, while the rift against the blackness of the cloud makes the flash'.[10] The word *pneuma*, which may also be translated as 'a blowing', carries several meanings, including wind, breath, spirit and air; the word *anemos* also refers to 'wind', but does not carry such a complex array of possible readings. Seneca is also a source here, stating that 'Anaximander referred everything to *spiritus* [*ad spiritum*]: thunder he said, is the noise of smitten cloud . . .'.[11] Here, the choice of the word *spiritus* reflects a similar semantic complexity to *pneuma*, the Latin word *ventus* being closer in meaning to *anemos*. *Pneuma/spiritus* (blowing, breath, wind, air, spirit) offers a greater semantic range and richer sense of complexity than does *anemos/ventus* (wind). It would be interesting to know whether Theophrastus was the original source for this emphasis on *pneuma* attributed to Anaxi-mander. Wind appears to have been a topic of particular interest to Theophrastus, who wrote a separate treatise on the subject.

Anaximenes of Miletus (fl. 546–525 BCE) was said to have been a student of Anaximander;[12] Aristotle, Theophrastus and Aëtius all attest to his interest in meteorological phenomena. In the *Metaphysics*, Aristotle briefly noted that for Anaximenes air (*aēr*) was the primary form of matter,[13] although it is not entirely certain how to understand the term *aēr* here.[14] Simplicius (sixth century CE), in his commentary on Aristotle's *Physics*, quoted Theophrastus' explanation of Anaximenes' view that air (*aēr*) undergoes changes through which it takes on other material forms, as it either condenses or becomes more rarefied. In this way, air provides the basis of other forms of matter, including clouds and wind. So, 'being made finer [air] becomes fire, being made thicker it becomes wind, then cloud, then (when thickened still more) water, then earth, then stones; and the rest come into being from these'.[15] The changes undergone by air through condensation and rarefaction are crucial to explaining weather. Aëtius reported that 'Anaximenes said that clouds occur when the air is further thickened; when it is compressed further, rain is squeezed out, and hail occurs when the descending water coalesces, snow when some windy portion is included together with the moisture'.[16]

There is a strong sense of a continuum in Anaximenes' explanations, at least as they are reported by others. Air was the primary form of matter, changing according to condensation and rarefaction. The stuff of the atmosphere is the same stuff as the earth; wind is related to stone. Following this line of thought, the inclusion of earthquakes as topics of meteorological discussion along with wind and rain (water) makes sense, for all of these phenomena share an underlying material basis.[17] The overarching primacy of air in Anaximenes' natural philosophy is emphasized. Indeed, Aëtius suggests that for Anaximenes air (as wind or breath) encloses the entire world, in a way similar to that in which the soul, also composed of air, holds together and controls humans:

> Anaximenes son of Eurystratus, of Miletus, declared that air [*aēr*] is the principle of existing things; for from it all things come-to-be and into it they are again dissolved. As our soul, he says being air [*aēr*] holds us together and controls us, so does wind [*pneuma*] (*or* breath) and air enclose the whole world. (Air and wind are synonymous here.)[18]

There are difficulties in understanding this passage, and doubts regarding the reliability of Aëtius' quotation of Anaximenes.

Nevertheless, the suggestion that wind functions in the world in a manner somehow similar to the air (or breath) inside humans is very striking, and is a suggestion which recurs, in various ways, in other meteorological accounts. The comment that 'air' and 'wind' are synonymous is presumably Aëtius' gloss on Anaximenes' views, but it is difficult to know how much of Anaximenes' own views have been preserved and communicated.[19]

Natural philosophers are not the only sources of knowledge about ancient meteorology. Aristophanes (born *c.* 447–445, died between 386 and 380 BCE), in his play *The Clouds*, provides a humorous view, presented largely through the character named 'Socrates'. Diogenes of Apollonia (fl. 440–430 BCE) is often assumed to be a model for some of the speeches given by this 'Socrates'. He may have been a physician; Aristotle reports his theory of *phlebes* (veins).[20] Simplicius credits Diogenes with having written a (now lost) book on meteorology.[21] Theophrastus (quoted by Simplicius) describes him as an eclectic writer, whose ideas are linked to a number of other Presocratics;[22] Diogenes' notion of the primacy of air may be traced to Anaximenes.

We may get some sense of Aristophanes' own view of the silliness of meteorological investigation from Socrates' speech in the *Clouds* (227–34):

> Why, for accurate investigation of meteorological phenomena it is essential to get one's thoughts into a state of, er, suspension by mixing small quantities of them with air – for air, you know, is of very similar physical constitution to thought – at least, to mine. So I could never make any discoveries by looking up from the ground – there is a powerful attractive force between the earth and the moisture contained in thought. Something similar may be observed to happen in the case of watercress.[23]

There is a strong impression here that those who investigate *ta meteōra* are 'airheads'. In using comedy to criticize the new learning of Socrates and other 'sophists' who, he suggested, were 'always talking about clouds and things', Aristophanes provides amusing evidence of the interests of some ancient Greeks in theorizing about meteorological phenomena. Diogenes of Apollonia should not be understood as the only target of Aristophanes' barbs; if meteorological theorizing had been limited to only a few people, the point of the parody would have been lost on his audience.

Aristotle

Aristotle was born at roughly the time that Aristophanes died. Aristotle makes it clear that his *Meteorology* is part of his broader programme to investigate the natural world.[24] The range of his writings is enormous, and not restricted to the natural world, encompassing such disparate topics as the constitution of Athens, the parts of animals, language, logic, poetics, ethics and theology, to name only some of the subjects covered in his works.

In the opening of the *Meteorology*,[25] Aristotle indicates that it is part of the larger project. The *Meteorology* is tied to a wider investigation which includes the study of nature and natural motion begun in the *Physics* and continued in his work on astronomy, *On the Heavens*, where he deals with the ordered movements of the stars. Aristotle continues his project in *On Coming-to-be and Passing-away* (also known as *On Generation and Corruption*), where he considers the number, kinds and changes of the four elements (Fire, Air, Water and Earth), as well as growth and decay in general. Aristotle announces that once he has completed the *Meteorology*, he will turn his attention to animals and plants. Within this hierarchical treatment of nature – working from the top down, so to speak – meteorology is in the middle. This central position reflects the mediating role that meteorological processes serve; meteorological phenomena are caused by the motions of the celestial region and, in turn, affect living things on earth.

At the beginning of the *Meteorology*, Aristotle defines the topic of inquiry as 'everything which happens naturally, but with a regularity less than that of the primary element of bodies, in the region which borders most nearly on the movements of the stars'. Aristotle then lists examples of meteorological phenomena that he will discuss, including the Milky Way, meteors, earthquakes and thunder. Part of the definition of the subject-matter involves the location in which meteorological phenomena occur, that is, below the region of celestial motion. But this subject-matter is not defined only by location; it includes 'all phenomena that may be regarded as common to air and water, and the various kinds and parts of the earth and their characteristics'.[26] Accordingly, Aristotle covers the following topics in the *Meteorology*: shooting stars, colourful phenomena at night (which may include the aurora borealis),[27] comets, rain, clouds, mist, dew, snow, hail, rivers and springs, coastal erosion and silting, the origin, place and saltiness of the sea, thunder and lightning, hurricanes, typhoons, *prēstēr* (whirlwinds)[28] and thunderbolts, haloes, rainbows, 'rods' (presumably, 'sun pillars') and parhelia ('mock suns'). Meteorology also includes 'the investigation of the causes of winds and earthquakes and all the occurrences associated with their motions'.[29] From the modern perspective these phenomena are not all meteorological; they are not all atmospheric or weather phenomena. But understanding their relationship in Aristotle's terms is crucial; for him, they are all liable to explanation through a common explanatory principle. Briefly, they are the 'products' of exhalations and involve changes from one terrestrial 'element' to another.

Explanation

Aristotle was keenly aware of the challenges of offering explanations of natural phenomena. Near the beginning of the *Meteorology*, he warns us (339a2–3) that, with regard to meteorological phenomena, in some cases we are puzzled, others we comprehend in a way. In other words, some of what will be covered will be difficult (possibly impossible) to grasp, while some will be understandable to some extent. The reader has been warned: the study of meteorology is not straightforward, and certain knowledge cannot be guaranteed. Whether or not the topics considered are truly inexplicable or 'merely' difficult is an important question. At the beginning of his treatise *On the Soul* (402a10–11), Aristotle issues a similar caveat, that 'to attain knowledge about the soul is one of the most difficult things in the world' and then turns to a discussion about methodology. However, in the case of the *Meteorology*, no such focused treatment of method directly follows the warning. Rather, the caution regarding the difficulty of the subject occurs within a list of phenomena to be investigated, reinforcing the sense of tentativeness.

To some extent, the special difficulty involved in explaining meteorological phenomena may be due to the problem of accessing information about them. Factors such as distance and difficulty of observation (for example, in the case of clouds) and rarity of occurrence (in the case of lunar rainbows, see *Meteorology* 372a28f.) limit the ability to observe the phenomena. The sense of tentativeness and lack of confidence in accessing phenomena is reinforced in Aristotle's statement of his aims in the *Meteorology*. He explains that 'we think that we have given a rational enough account of things not apparent to the senses when we have produced a theory which leads back to what is possible'.[30]

Aristotle's proclaimed satisfaction with a 'rational enough account' of 'what is possible' may seem surprising in light of some of his discussions of the nature of knowledge, understanding and explanation in other works, specifically the *Metaphysics* and the *Posterior Analytics*. For Aristotle, knowledge consists of *epistēmē*, understanding. Epistemic knowledge (sometimes translated as 'science') is a body of systematically arranged information, in the form of proofs or demonstrations, involving deductions from first principles. As Aristotle explains in *Posterior Analytics* (1.2), scientific knowledge is the knowledge of causes. For him, to know something is 'to know the cause or reason why it must be as it is and cannot be otherwise'.[31] His overall project is to provide explanations and understanding,[32] yet, from the outset of the *Meteorology* he makes it clear that this understanding will be limited.

In the *Posterior Analytics*, Aristotle presents a formal model of scientific knowledge (*epistēmē*); such knowledge relies on demonstration (*apodeixis*). Demonstration depends on the use of deductive inference, in the form of a *sullogismos*. Elsewhere, in the *Prior Analytics*, Aristotle defines a *sullogismos* as 'an argument in which, certain things being assumed, something different from the things assumed follows from necessity by the fact that they hold'. An example is 'If *A* is predicated of every *B*, and *B* of every *C*, *A* must be predicated of every *C*'.[33]

Even though Aristotle warns us that the subject of meteorology is 'difficult', he includes some meteorological examples in the *Posterior Analytics*, in his discussion of scientific explanation. Such explanations require a definition that links what is to be explained (the explanandum) to pre-eminently knowable first principles, by means of a deductive framework.[34] Aristotle illustrates the distinction between definition and demonstration by the following: 'What is thunder? Extinction of fire in cloud. Why does it thunder? Because the fire in the cloud is extinguished.'[35]

But many of Aristotle's treatises, including the *Physics*, *On the Heavens*, *Generation of Animals*, *Parts of Animals* and *Meteorology* do not obviously conform to the model laid out in the *Posterior Analytics*; the explanations offered in these works are not generally presented syllogistically. As Cynthia Freeland has noted, 'notoriously, Aristotle's scientific treatises do not exhibit the syllogisms we would expect of ideal sciences. Instead, in these treatises causal explanations are typically offered and clearly marked according to category or type. It might well be true, then, that in scientific practice . . . [Aristotle] is utilizing a notion of explanation distinct from that set forth in the *Posterior Analytics*.' The question of the relationship between the formal method laid out in the *Posterior Analytics* and the explanations offered in Aristotle's scientific treatises continues to be the subject of heated debate, which cannot be rehearsed here. [36]

In the *Physics*, Aristotle explains that there are four kinds of causes, the material, formal, efficient and final; in order to offer an explanation of something, it is necessary to identify its causes. He briefly outlines the four causes as follows:

> In one way, then, that out of which a thing comes to be and which persists, is called a cause, e.g. the bronze of the statue, the silver of the bowl, and the genera of which the bronze and the silver are species.
>
> In another way, the form or the archetype, i.e. the definition of the essence, and its genera, are called causes (e.g. of the octave the relation of 2:1, and generally number), and the parts in the definition.
>
> Again, the primary source of the change or rest; e.g. the man who deliberated is a cause, the father is cause of the child, and generally what makes of what is made and what changes of what is changed.
>
> Again, in the sense of end or that for the sake of which a thing is done, e.g. health is the cause of walking about. ('Why is he walking about?' We say: 'To be healthy', and, having said that, we think we have assigned the cause.)[37]

These four causes can be understood as providing the answers to four different sorts of questions: 'What is it composed of?', 'What is it?', 'What brought it about?' and 'What is it for?'[38] While the provision of the four causes may represent an ideal explanation for Aristotle, some of his writings, including the *Meteorology*, do not always conform to this model.

In the *Meteorology*, Aristotle is particularly concerned with the material and efficient causes of the phenomena. At the very beginning of the work, he states that the elements themselves (Fire, Air, Earth and Water) are material causes, and the motion of the ever-moving bodies (in the celestial sphere) should be regarded as the efficient cause of terrestrial events.[39] In other words, the four terrestrial elements are what the *meteōra* are composed of, while the celestial motions have brought the *meteōra* about. So, for example, in his discussion of rain, Aristotle states that the circle of the sun's revolution is the moving (efficient) cause.[40]

But it is Aristotle's concern with knowledge of final causes (teleology) that is often regarded as characteristic of his natural philosophy. Particularly in his studies of living things, Aristotle points to the final cause, the reason why things are as they are, that is, the goal for which they have come to be.[41] Yet in spite of his commitment to the importance of purposiveness, Aristotle does not always state the final cause of particular phenomena; in the *Generation of Animals* he makes it clear that there are some things in nature which seem to have no purpose (some parts of animals, for example).[42]

Within the *Meteorology*, Aristotle is not concerned with spelling out the final cause. But, elsewhere, in the *Physics* he discusses the purpose of rainfall; this discussion is part of his larger argument about the purposefulness of natural processes. First he raises the question:

> why should not nature work, not for the sake of something, nor because it is better so, but just as [Zeus] rains, not in order to make the corn grow, but of necessity? (What is drawn up must cool, and what has been cooled must become water and descend, the result of this being that the corn grows.) Similarly, if a man's crop is spoiled on the threshing-floor, the rain did not fall for the sake of this – in order that the crop might be spoiled – but that result just followed. [43]

Here, Aristotle puts forward a possible opponent's view, which offers an explanation of rain that incorporates both a traditional component (Zeus), and a physical cause (water as it is heated by the sun rises, then cools and condenses and falls to earth as rain).[44] The juxta-position of these two explanatory elements (traditional god and physical cause) may be Aristotle's shorthand way of pointing to efforts to shift the understanding of rain as an act of a god to a purely natural process.[45]

But the physical explanation offered in this passage corresponds closely to Aristotle's own descriptions of the physical process that produces rain. For example, in *On Sleep*, as part of a discussion of cooling effects in nature (with reference to the brain), Aristotle briefly explains rainfall this way: 'moisture turned into vapour by the sun's heat is, when it has ascended to the upper regions, cooled by the coldness of the latter, and becoming condensed, is carried downwards, and turned into water once more'.[46] In the *Posterior Analytics*, rain is offered as an example of the sort of 'circular' coming-about: 'if the earth is soaked, necessarily steam came about; and if that came about, cloud; and if that came about, water: and if that came about, it is necessary for the earth to be soaked'.[47] A similarly brief explanation of rainfall is presented in *Parts of Animals* (653a2ff.); in the *Meteorology* (346b21ff.), the same explanation, only slightly elaborated, is given. So, there is ample evidence that Aristotle is willing to endorse a physical explanation of rainfall.

But, in the *Physics*, Aristotle emphasizes the teleological and purposeful nature of rain:

> all . . . natural things either invariably or for the most part come about in a given way; but of not one of the results of chance or spontaneity is this true. We do not ascribe to chance or mere coincidence the frequency of rain in winter, but frequent rain in summer we do; nor heat in summer but only if we have it in winter. If then, it is agreed that things are either the result of coincidence or for the sake of something, and these cannot be the result of coincidence or spontaneity, it follows that they must be for the sake of something; and that such things are all due to nature. . . . Therefore action for an end is present in things which come to be and are by nature.[48]

In this passage he also confronts how we are to explain the occurrence of rainfall in different seasons. He uses the example of rain occurring in different seasons to illustrate his view that those things which happen by nature (and so are liable to scientific explanation involving causes) happen 'always or for the most part', and not by chance.[49] In the *Physics* this is part of his larger argument that 'nature belongs to the class of causes which acts for the sake of something', emphasizing his commitment to the teleology of natural processes. In the *Physics*, he is not particularly concerned with explaining rain; rather the explanation of rain serves as an example of a larger argument about nature. In the *Meteorology* (346b16ff.), Aristotle makes it explicit that he is focusing there on causes other than the final cause (the purpose) of rain; there he explains rain as being due to the efficient cause, the sun's motion as it gets closer to or more distant from the earth, and the subsequent processes of evaporation and condensation. But, while Aristotle is not concerned with final causes in the *Meteorology*, this does not mean that he does not ascribe a final cause to meteorological phenomena.

Elsewhere, in his discussions of scientific explanation (in the *Posterior Analytics* and the *Metaphysics*), Aristotle emphasizes that it is not the case that events occur invariably; as he puts it, some things happen only 'for the most part', others occur only rarely.[50] Aristotle addresses the issues regarding those things that happen 'for the most part' at several points. In the *Metaphysics* (1027a20–6) he explains that:

> all science is either of that which is always or of that which is for the most part. For how else is one to learn or to teach another? The thing must be determined as occurring either always or for the most part, e.g., that honey-water is useful for a patient in a fever is true for the most part. But one will not be able to state when that which is contrary to this happens, e.g. 'on the day of the new moon'; for then it will be so on the day of new moon either always or for the most part.[51]

This passage is part of a larger discussion denying the possibility of a science (i.e. knowledge derived through demonstration) of the accidental. For Aristotle, those things that happen by chance are not capable of a proper explanation. Although an account can be given of such things, they lack a purpose (or final cause). But this does not mean they can be totally ignored; in some cases, Aristotle explains, 'we should also base our arguments upon what happens for the most part as well as upon what necessarily happens'.[52]

This is particularly relevant to the discussion of rainfall, which according to Aristotle, happens 'naturally' at a certain time of year (winter), but occurs by chance or coincidence during the summer. In the *Physics*, Aristotle's explanation of rainfall seems to point to 'for the sake of the growth of plants' as the final cause: the reason why or purpose for which rain occurs by nature, as it does, 'always or for the most part'.[53] In his study of the natural world, Aristotle adheres to the view that the kinds of living things which exist in the world are permanently preserved. The preservation of natural kinds is invoked as the aim, or goal, towards which the final cause is directed.

Working from this assumption of the preservation of kinds of living things, the study of nature includes the investigation of natural phenomena that contribute to that preservation.[54] Aristotle's consideration of seasonal rainfall in the *Physics* may be his only explicitly teleological treatment of such phenomena. In the *Meteorology*, he restricts his discussion to material and efficient causes. It may be that he regarded his compartmentalized study of meteorological phenomena as the wrong place to address questions such as the purpose for which rain falls, because the reason for rainfall involves living things not treated in the *Meteorology*. David Sedley has suggested that the reason why the final cause of rain (and by extension other phenomena) is not addressed in the *Meteorology* is because the final cause can be properly understood only 'when the world's interactive structure is examined as a whole, from the top down'. It is only then that 'the contribution of weather, animals and everything else to the natural hierarchy moves into focus'.[55] This interpretation seems reasonable, and is underscored by the place of the *Meteorology* in the middle of the hierarchical study of nature that Aristotle outlined in the beginning of the work. He does not appear to have been specifically interested in the explanation of meteorological phenomena but, as part of the larger project of natural philosophy, meteorological processes have to be explained. That the full explanation of these processes and phenomena is not offered in the one work, the *Meteorology*, is an indication that he was not aiming to explain everything about the *meteōra* in this treatise. Rather, for Aristotle, a full understanding of meteorology can be attempted only with reference to the study of other aspects of the natural world.

So it should be no surprise that some of Aristotle's other writings provide useful insights for understanding his definition of meteorology and the inclusion of particular types of phenomena. In fact, near the beginning of the *Meteorology* (in the second and third chapters of Book 1), Aristotle acknowledges this himself and so provides a brief summary discussion of his views as presented in *On the Heavens* and *On Generation and Corruption*.[56] The physical world is, for Aristotle, divided into two regions, the terrestrial and the celestial. Each region is characterized by distinct material elements and motions. Four elements are present in the terrestrial region: Earth, Water, Air and Fire. The natural motions of bodies in the terrestrial region are rectilinear: that is, either upwards, away from the centre of the universe, or downwards, towards the centre of the universe.[57] As Aristotle himself explains in the *Meteorology*, he had earlier (in other works, including *On the Heavens*) discussed the existence of a single element that composes the celestial bodies whose natural motion is circular;[58] that element, in the celestial region, is known as the *aithēr*.[59]

Aristotle explains that there are four other bodies, Fire, Air, Water and Earth, which are differentiated by the primary contrary qualities, unnamed in the *Meteorology* but identified elsewhere as hot, cold, dry and moist.[60] These four physical bodies move either away from or towards the centre of the universe; Fire and Air move away from the centre, while Earth and Water move towards the centre.[61] These four bodies (or elements) exist in comparable quantities to one another[62] and, furthermore, they change one into the other.[63] The physical bodies encountered in everyday life are not 'pure' forms of the elements, but are 'compounds' (containing at least two of the elements); so, for example, what we know as everyday, garden variety 'earth' is a compound composed of the elements Earth, Water, Fire and Air.[64] The *Meteorology* is concerned with changes that affect these bodies.

Aristotle discusses the types of matter (and related motions) present in the different regions of the universe, and asks (*Meteorology* 339b13–15) 'are we to consider that one physical substance occupies the space between the earth and the farthest stars, or more than one?' He sets up and knocks down two straw men, arguing against the views of unnamed others who suggest that the region is full of either air or fire. Aristotle argues that the region from the surface of the earth up to the celestial realm is occupied neither by air or fire alone, stating his own views as follows (340b6–10): 'the upper region as far [down] as the moon we affirm to consist of a body distinct both from fire and from air, but varying in degree of purity and in kind, especially towards its limit on the side of the air, and of the world surrounding the earth'. Aristotle suggests that the *aithēr*, the material substance of the celestial region, varies in purity, being more pure where it is more distant from the terrestrial elements, and less pure in closer proximity to them.[65] But it is not the material composition of the *aithēr* but rather the effects of the motions of this substance which concern us. Aristotle explains that 'the circular motion of the first element and of the bodies it contains dissolves, and inflames by its motion, whatever part of the lower world is nearest to it, and so generates heat'.[66] In other words, the motion of the *aithēr* causes material in the terrestrial realm to be heated.

Having briefly reminded us of the four primary forms of terrestrial matter, and their natural places within the cosmos, Aristotle is able to return to the question of the nature of 'what we call the air' in the terrestrial region, explaining that (*Meteorology* 340b24ff.):

> we must understand that of what we call air the part which immediately surrounds the earth is moist and hot because it is vaporous and contains exhalations from the earth, but that the part above this is hot and dry. For vapour is naturally moist and cold and exhalation hot and dry: and vapour is potentially like water, exhalation like fire. . . . For the whole encircling mass of air must necessarily be in motion, except that part of it which is contained within the circumference

that makes the earth a perfect sphere. It moves in a circle because it is carried round by the motion of the heavens. For fire is contiguous with the element in the celestial regions, and air contiguous with fire, and their movement prevents any condensation.[67]

Even though it resides in the terrestrial region, air is affected and carried around by the motions of the celestial region.

Here, in the *Meteorology*, Aristotle presents a much fuller account of interactions between the celestial and the terrestrial regions than he did in the *Physics* and in *On the Heavens*. In the latter work, Aristotle explains that the heat and light of the stars, including the sun, is the result of their motion creating friction which causes the air below to be ignited; the celestial material itself is neither hot nor cold. Locomotion, that is, change of place, is the only type of change that occurs in the celestial realm, and it is the celestial motion that produces heat. This is the reason we feel hot when the sun is near or overhead, because the air is especially heated. To account for this, Aristotle attributes a change in the terrestrial region to the motion of the celestial bodies.[68] Similarly, in the *Meteorology* (340b12–14), Aristotle explains that heat which is felt in the terrestrial region is caused by celestial motions.

Aristotle offers two explanations of how the terrestrial region is heated by motions within the celestial region. First (at *Meteorology* 341a13ff.), he explains the heat generated by the sun itself.[69] This heat plays a crucial role in his meteorological theory, for it is from heat generated by the sun that two types of exhalations rise up from the earth, and contribute to various meteorological phenomena. According to Aristotle, the celestial bodies are not themselves hot. However, motion can inflame and rarefy air. He argues that the proximity of the sun and its motion are particularly conducive to warming in the terrestrial region. The stars, while they move quickly enough to produce heat, are not close enough; the moon is closer, but moves too slowly. Only the sun has both the necessary speed and proximity to produce heat in the terrestrial region: 'the sun's motion is therefore in itself sufficient to produce warmth and heat: for to produce heat a motion must be rapid and not far off'. Aristotle (341a29–32) very briefly offers a second reason for the heat caused by celestial motions: the fire which surrounds the air of the terrestrial region is frequently scattered by the motion of the heavens and carried forcibly downwards.[70] But he is at pains to argue that the celestial region itself is not hot or fiery, and he offers as supporting 'evidence'

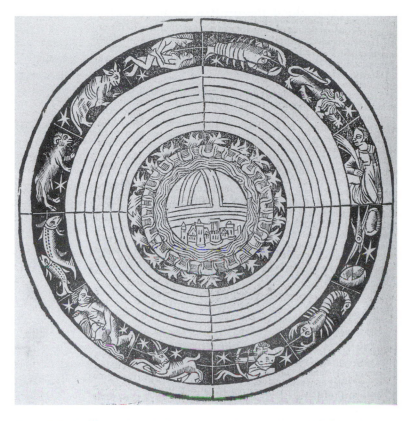

Figure 3.1 An illustration of the universe, showing Aristotle's four elements (Earth, Water, Air and Fire in order outwards from the centre), from Ramón Lull (d. 1315), *Practica Compendiosa artis Raymundi Lull* (1523), by permission of the Syndics of Cambridge University Library.

Source: Cambridge University Library.

(σημεῖον) his own view that shooting stars originate in the terrestrial region.

While there are two distinct regions in Aristotle's cosmos, the celestial and the terrestrial, which are distinguished by their different material substances, these two regions do interact. In his view, celestial motions cause material in the terrestrial realm to be heated. In the case of meteorology, these interactions are crucial, for they are the underlying causes of many phenomena. Aristotle (*Meteorology* 341b13–24) explains that:

the region round the earth is arranged as follows: first, immediately beneath the circular celestial motion comes a warm and dry substance which we call fire (for we have no common name to cover every subspecies of the smoky [καπνώδους] exhalation: but because it is the most inflammable of all substances, we must adopt this nomenclature); below this substance comes air. Now we must think of the substance we have just called fire as extending round the outside of the terrestrial sphere like a kind of inflammable material, which often needs only a little motion to make it burst into flames, like smoke: for flame is the boiling up of a dry current of air. Wherever then conditions are most favourable this composition bursts into flames when the celestial revolution sets it in motion.[71]

In this passage, it is clear that the effect of the celestial motion on the material terrestrial region is fundamental to Aristotle's exposition of meteorology.

According to him, the topic of his study was called meteorology by 'all our predecessors'. Yet it is not clear that his definition of the subject was that of his predecessors, which he approves and adopts, or whether it is his own newly fashioned definition, loosely based on those of earlier thinkers. While his direct predecessor and teacher Plato seems to have defined the *meteōra* as lofty things,[72] some of the meteorological phenomena described and explained by Aristotle were far from lofty, being located in the depths of the earth. For him, meteorology is concerned with subterranean events as well as things located higher up. So, for example, wind plays a very important role in his meteorological theory, as a cause of earthquakes. While Aristotle mentions some of his predecessors' views on earthquakes (including those of Anaximenes, Anaxagoras and Democritus), it is not clear to what extent they considered phenomena such as earthquakes to be meteorological, or whether Aristotle's inclusion of them in a discussion of the *meteōra* was an innovation, related to his (in his own view) original theory involving two exhalations.[73]

Exhalations

In the *Meteorology*, most of the phenomena are explained in terms of 'exhalations'. These exhalations are crucial to Aristotle's meteorology, for they provide the material cause of the *meteōra* (342a28–9); he contrasts his own theory involving two exhalations with those of (unnamed) others who posited only one exhalation. Aristotle mentions the exhalations at several points, but does not offer a single unified statement defining and describing them. In order to understand the exhalations, it is necessary to piece together scattered references to their composition and action.

Aristotle first touches on the topic in Book 1, where he introduces the exhalations as part of his discussion of the composition of the region surrounding the earth. The two different 'exhalations' from the earth are invoked to explain most meteorological phenomena (including those that are 'atmospheric' as well as others which are 'astronomical' and 'geological', in modern terminology). These exhalations (sg. = ἀναθυμίασις) which

> arise from the earth when it is heated by the sun must be not, as some think, of a single kind, but of two kinds; one is more vaporous in character (ἀτμιδωδεστέραν), the other more windy (πνευματωδεστέραν), the vapour arising from the water within and upon the earth, while the exhalations from the earth itself, which is dry, are more like smoke. The windy exhalation being hot rises to the top, the more watery exhalation being heavy sinks below it.[74]

At several points in the *Meteorology*, Aristotle reiterates that his theory is based on the existence of two exhalations, in some cases providing specifics relating to particular phenomena (such as the saltiness of the sea, 2.3, 357b24–6). Rather far into the work (2.4), in his discussion of the wind, he notes that the definition and description of the exhalations are not straightforward:

> we have said that there are two kinds of exhalation – one moist and one dry: of these the first is called vapour (ἀτμίς), the second has no name that applies to it as a whole, and we are compelled to apply to the whole a name which belongs to a part only and call it a kind of smoke. The moist exhalation does not exist without the dry nor the dry without the moist, but we speak of them as dry or moist according as either quality predominates.[75]

As he explains, sometimes the vaporous exhalation dominates, at other times the dry one (360b2–4). But, in spite of the difficulties in describing the exhalations, Aristotle is nevertheless clear that their role is fundamental to meteorology, providing the material cause.[76]

In addition to describing the exhalations, Aristotle outlines the processes which produce various meteorological phenomena. The place of their formation is important. So, for example, in his discussion of burning flames, shooting stars, torches and 'goats', some are formed in the upper portion of the terrestrial region; when the (dry) exhalation is ignited by celestial motions, phenomena such as shooting stars may be produced.[77] When these phenomena are formed in the lower part of the terrestrial region, this is due to ejection because the more humid exhalation has been condensed and cooled. The exhalations provide the material cause of such phenomena, while either the celestial motion or the condensation of the air as it contracts serves as the moving (efficient) cause. Aristotle argues that such phenomena as shooting stars occur below the sphere of the moon, and suggests that their speed is comparable to that of objects thrown by people. Such objects only 'seem to move much faster than the stars and sun and moon because they are close to us'.[78]

The two exhalations are the cause not only of such 'lofty' meteorological phenomena as comets and lightning, but also of *meteōra* lower down. So, for example, the winds are exhalations whose movement is caused by celestial motions.[79] Since the wind is

> a body of dry exhalation moving about the earth, it is clear that though their motion takes its origin from above the material from which they are produced comes from below. Thus the direction of flow of the rising exhalation is determined from above, as the motion of the heavens controls things whose distance from the earth is considerable: at the same time the exhalation rises vertically from below, since any cause operates more strongly on its effect the nearer it is to it and the exhalation is clearly produced originally from the earth.[80]

Aristotle adds that 'the facts also make it clear that winds are formed by the gradual collection of small quantities of exhalation, in the same way that rivers form when the earth is wet'.[81] Exhalations are also responsible for earthquakes, which are due to trapped exhalations that cannot escape from within the earth. Other meteorological phenomena, including thunder, lightning, whirlwinds, firewinds, thunderbolts and hurricanes, are also due to the forced ejection of the dry exhalation from within clouds.[82]

At the end of Book 3 of the *Meteorology*, Aristotle sums up the topics that have been covered, stating that he has completed the 'enumeration of the effects produced by exhalation in the regions above the earth's surface', emphasizing that all of the phenomena that have been discussed up to that point (including comets, thunder, lightning, rain, snow, hail, winds, earthquakes, haloes and rainbows) are due to exhalations. He states that there are other effects produced by exhalations enclosed within the earth, namely, quarried substances and metals.[83]

Spatially, most meteorological phenomena occur in the region located physically between the celestial sphere and the surface of the earth (but remember that earthquakes are meteorological phenomena as well). The subject-matter of the *Meteorology* occupies an intermediate, bridging position between the two realms of Aristotle's cosmos, and not only spatially. While Aristotle may have a formal scientific method which he describes in the *Posterior Analytics*, in his works on nature he often employs a variety of 'informal' approaches and strategies to describe and explain phenomena. Meteorology also appears to have an intermediary position, methodologically, in relation to other branches of the study of the natural world, including astronomy, physics and the study of living things. This intermediary position is hinted at in the introduction to the *Meteorology*. There Aristotle states the conceptual and hierarchical order of his studies on nature, where considerations of motion and the elements take precedence over his treatment of animals and plants.[84] In the *Meteorology*, Aristotle employs a number of approaches to describe and explain meteorological phenomena.[85] The idea that meteorology is an intermediary study is borne out by the variety and mix of explanatory strategies employed, including the use of analogy (often utilized as well in discussions of living things, e.g. in *Parts of Animals*)[86] and geometrical diagrams (also used, for example, in astronomy, e.g. *On the Heavens* 1.5).

In the *Metaphysics*, Aristotle outlines three different types of theoretical knowledge, with three different subject areas, namely, mathematics, natural science (physics) and theology.[87] In the *Posterior Analytics*, Aristotle indicates that various kinds of theoretical knowledge operate independently; so 'one cannot therefore prove anything by crossing from another genus – e.g. something geometrical by arithmetic'.[88] He states that 'the reason why differs from the fact . . . when each is considered by means of a different science'. He offers the following example of the sort of information offered by different (and unrelated) approaches: 'it is for the doctor to know the fact that circular wounds heal more slowly, and for the geometer to know the reason why'.[89] Aristotle specifically addresses one of the topics treated in the *Meteorology*, that is, the rainbow. In the *Posterior Analytics*, he explains that there is a sort of division of labour involved in studying the rainbow. Thus, 'it is for the natural scientist to know that fact, and for the student of optics . . . the reason why'.[90] While there are issues about the relationship between the physical and non-physical sciences, particularly with regard to the apparent prohibition against kind-crossing (traversing the 'boundaries' between different branches of knowledge), it is notable that in the *Meteorology* some of the explanatory strategies employed are common to other areas of natural philosophy, while others are more specifically mathematical. Furthermore, some explanatory approaches are not restricted to the specialist, but are part of common experience.[91]

Sources of information

As Aristotle warned early in the *Meteorology*, the study of the *meteōra* is difficult, and will not yield complete understanding. To aid in his study, he gathers together information about meteorological phenomena from a range of sources; in his discussion of the causes of these phenomena, information and 'facts' are presented in a variety of ways.

Aristotle is interested, in the *Meteorology* as in his other writings on the natural world, in providing a general account; he does not, for the most part, focus on individual cases. As R.J. Hankinson has explained, Aristotle 'is not really concerned with the particular causes of individual events, but with the general patterns which run invariably (or at least for the most part) through the structure of the world'.[92] But Aristotle does, at times, make reference to specific meteorological incidents, for example earthquakes and tidal waves. So, for example, he mentions (368b6–7) a tidal wave in Achaea, which occurred at the same time as an earthquake (in 373–372 BCE). Aristotle himself would have been eleven years old at the time; he refers to these events in a way which suggests that these meteorological disasters would have been well known to his audience.[93]

In some of his work, notably his writings on living things and in his collection of political constitutions, Aristotle collated information from a variety of resources. Similarly, in the *Meteorology*, Aristotle brings together data and examples derived from a number of sources. Within his meteorological explanations, he intersperses specific details about conditions in different places; it is likely that he collected this information from various informants. So, for example, in his discussion of the nature of salt water, he states:

The following facts also all support our contention that it is
the presence of a substance that makes water salt, and that
the substance present is earthy. In Chaonia there is a spring
of brackish water which flows into a neighbouring river that
is sweet but contains no fish. For the inhabitants have a story
that when Heracles, on his way through with the oxen from
Erytheia, gave them the choice, they chose to get salt instead
of fish from the spring. For they boil off some water from it
and let the rest stand; and when it has cooled and the moisture has evaporated with the heat salt is left, not in lumps
but in a loose powder like snow. It is also rather weaker than
other salt and more of it must be used for seasoning, nor is
it quite so white. Something of a similar sort happens also
in Umbria. There is a place there where reeds and rushes
grow: these they burn and throw their ashes into water and
boil it till there is only a little left, and this when allowed
to cool produces quite a quantity of salt.[94]

While it is not clear whether Aristotle gathered this information
himself or relied on the experiences and accounts of others, the speci-
ficity of his details provides a certain homeliness reminiscent of
Herodotus. Various people, in different times and places, might have
been persuaded by the credibility of the reports, while others may
have been convinced by the authority of Aristotle.

While many of the meteorological phenomena described are
common occurrences, others are more rare. So, for example, in his discussion of the rainbows that occur at night, Aristotle explains that
'the ancients' did not think such rainbows actually occurred; their
disbelief was due to the rarity of the event. Aristotle noted
(*Meteorology* 372a22ff.) that he had met with only two examples of
the night rainbow in more than fifty years. In some cases Aristotle
must have relied on second- or third-hand reports; these could not
have been entirely reliable.[95]

The role of endoxa

Aristotle does not claim to have collected all the information and data
reported in the *Meteorology* himself; he makes it clear he is relying on
information and reports. For specific rarely occurring and unusual
phenomena, he would have had to use information gathered and
reported by other observers. In other works, for example the *History
of Animals*, Aristotle sometimes uses such reports. The *Meteorology*

contains an accumulation of information collected from earlier natural philosophers, poets and from shared experience.[96] But he does not only present empirical information so derived. At various points throughout the *Meteorology*, he reports (and sometimes criticizes) ideas, theories and explanations offered by his predecessors and contemporaries. He does not always give the names of those whose views he reports. For example, in his discussion of the sea, Aristotle mentions the ideas of two groups of thinkers, theologians and philosophers, without attributing them to specific individuals. In discussions of some phenomena, other people are mentioned by name, including Anaximenes (fl. 546–525 BCE, on earthquakes), Anaxagoras (probably 500–428 BCE, on comets, the Milky Way, hail, earthquakes and lightning), Empedocles (*c.* 492–434 BCE, on the sea and lightning), the 'so-called Pythagoreans' (as he refers to them; on comets and the Milky Way), Democritus (b. 460–457 BCE, on comets, the Milky Way and earthquakes), as well as Hippocrates of Chios (fl. end of fifth century BCE) and his student Aeschylus (on comets). He often introduces the ideas of others in order to argue against them. At one point, as he criticizes (nameless) others' views about the winds, he makes it clear that ingenious theories have to be rejected if they are false.[97]

Nevertheless, it is part of his presentational tactics in the *Meteorology* to survey the views of others, especially 'the wise'. Such views, worth considering, are known as *endoxa*, defined in the *Topics* (100b21–4) as 'what seems so to all or the majority or the wise'. (The adjectival form *endoxos* may carry the meaning 'reputable' or 'respectable'; the plural noun *endoxa* can be understood as 'reputable opinions'.[98])

In the *Meteorology*, Aristotle does not usually accept the views of his predecessors, even when they are those of 'the majority or the wise'. It was not his aim simply to present a compilation of earlier ideas, although in cases such as his discussion of earthquakes he specifically mentions that three different theories have been offered prior to his time.[99] The discussion of others' views is part of the presentation of his own ideas. As in his other works, he sometimes introduces the opinions of others to serve as straw men, to be knocked down and discarded in favour of his own arguments. In some cases he labels the theories of others as not only false, but as 'silly' or 'absurd'; he goes so far as to call some 'childish'.[100]

Aristotle's use of *endoxa* as a starting point in his examination of particular phenomena may provide him with a rhetorical advantage, as he rejects the opinions of others. However, the roles that Aristotle's predecessors play in his presentation of their views is not as simple as it may seem. His motive in reviewing the ideas of others may not have been simply to refute them. Freeland argues persuasively that Aristotle's discussion of the *endoxa* presents a picture of natural philosophy as a 'problem-solving activity', in which important questions and problems are highlighted. Questions and problems arise from the process of critical examination of the *endoxa*, apart from the phenomena themselves. So, for example, Aristotle rejects Anaxagoras' explanation of hail, in which clouds are forced by the summer heat to the upper region, where the water in them freezes. He cites evidence, based on observations, to refute the explanation: Aristotle points out that hail falls from clouds close to the earth, rather than from far above, so contradicting Anaxagoras' claim.[101] Careful consideration of *endoxa* may actually motivate a programme of observation and data collection, in order to test (and if necessary contest) the views of others. So Aristotle presents surveys of his predecessors' *endoxa* on a variety of phenomena, including comets and hail.[102]

In reviewing the *endoxa*, Aristotle clearly engages with the ideas of others. His critical examination is serious (and not merely self-promoting) in that it helps him to develop an account that explains the phenomena. So, the engagement with the *endoxa* is a crucial step in his procedure in the *Meteorology*. Elsewhere, in the *Topics*, Aristotle defined 'the method by which we shall be able to reason from generally accepted opinions (ἔνδοξα) about any problem' as dialectic.[103] Dialectic, which proceeds from the starting point of reputable opinions, may be contrasted with the strict demonstrations of the *Posterior Analytics*, which proceed from self-evident, indemonstrable premises and aim at demonstrative certainty. That the explanation of meteorological phenomena does not proceed from such premises is no surprise; after all, we were warned at the beginning of the *Meteorology* that a full understanding of the phenomena cannot be offered. Explanations that build on *endoxa* may not yield certainty, but may offer, for certain subjects, the best answer available at a given time.[104]

Aristotle's discussion of the *endoxa* in the *Meteorology* may serve another purpose as well. By deliberately engaging with, utilizing and building upon the ideas of his predecessors and contemporaries, Aristotle presents himself as crucially dependent upon the work of others, even when that work is superseded by his own. His use of *endoxa* indicates that he regards science as a cumulative group enterprise,[105] in which the work of others in the community (predecessors and contemporaries) is shared and contributes to the larger effort to understand. Elsewhere, Aristotle makes it clear that this is his view of the manner in which understanding proceeds in other areas, including rhetoric and dialectic; he indicates that it is his (and others') usual practice to build on the work of predecessors.[106] Indeed, Aristotle does, in many instances, present himself as the culmination of the Greek project to explain natural phenomena.

This reinforces the picture of meteorological work presented in the last chapter, where it was shown that cumulative efforts over a long period had contributed to and were taken seriously by those interested in weather prediction. Aristotle presents himself as being the next in line of those interested in explaining meteorological phenomena.

Signs

Aristotle uses empirical observations in the *Meteorology*, particularly signs, as part of his programme to explain the phenomena. He makes it clear that he not only utilizes his own observations, but relies on those of others as well, particularly necessary in the case of rare phenomena, such as the moonbow (376b25). Freeland, as part of a more general discussion of the relationship between *endoxa* and signs, suggests that signs are 'particular sorts of *endoxa*, empirical facts available to *aisthesis*', that is, to sensation.[107]

In the *Prior Analytics*, Aristotle defines the term 'sign' in relation to its function within certain kinds of reasoning. A sign:

> means a demonstrative premiss which is necessary or gener-
> ally accepted. That which coexists with something else, or
> before or after whose happening something else has hap-
> pened is a sign of something's having happened or being.[108]

The question of what it means to be 'necessary' as opposed to 'gen-erally accepted' is important here, and is explained in the context of inferences further on in the passage. While some evidence is conclu-sive (and is called, in the singular, *tekmērion*), more often it is not wholly conclusive or sufficient for knowledge; such evidence is called a 'sign' (*sēmeion*).[109] Nevertheless, signs (*sēmeia*) serve to render a con-clusion a 'respectable or reputable thing (*endoxon*) to believe'.[110] From the context of Aristotle's discussion of the terms, it is clear that he is not so much concerned with their ordinary usage, as with their roles within inferences, and as they are used in explanations.[111] Significantly, Aristotle states that 'truth can be found in all signs'.[112]

Signs work as signs because there is some link, even if there is no universal or necessary connection.[113] Here, there is the explicit suggestion that, if conclusive evidence is not available, the 'next best' thing, a 'sign' or 'indication', will be useful and should be accepted.

In the *Meteorology*, Aristotle's discussion of meteorological 'signs' is part of his explanatory programme, and not part of a project aimed at weather prediction. His 'signs' are not presented in lists of phenomena which may indicate particular weather events, useful for prognostication. Rather, he uses 'signs' within the discussion and presentation of his meteorological theory, as evidence to support his explanation.

Earlier, Aristotle had explained the manner in which rain is formed:

> the moisture about it [the earth] is evaporated by the sun's rays and the other heat from above and rises upwards: but when the heat which caused it to rise leaves it, some being dispersed into the upper region, some being quenched by rising so high into the air above the earth, the vapour cools and condenses again as a result of the loss of heat and the height and turns from air into water: and having become water falls again onto the earth.[114]

As part of his discussion of haloes functioning as weather signs, Aristotle explains that, if the halo 'neither fades nor breaks, but is allowed to reach its full development, it is reasonable to regard it as a sign of rain, since it shows that a condensation is taking place of the kind, which, if the condensing process continues, will necessarily lead to rain'. He goes on to elaborate, explaining that a broken halo is a sign of wind, because its breaking up is due to a wind which has not yet arrived; 'an indication that this is so is that the wind springs from the quarter in which the main break occurs'. A faded halo is a sign of fine weather, for 'if the air is not yet in a state to overcome the heat contained in it and to develop into a watery condensation, it is clear that the vapour has not yet separated from the dry and fiery exhalation which causes fine weather'.[115] So his discussion of haloes as weather signs is specifically linked to his more general meteorological explanations, and is presented in the context of his meteorological theory, in which the exhalations play a central role. In this way, his discussion of the halo as weather sign serves to reinforce his claims regarding the superiority of his meteorological explanations; the 'signs' provide evidence for the theory. So, while the halo may indeed serve as an indication of impending weather, Aristotle is not here concerned with using weather signs for prediction. Rather, the traditional signs provide a special type of evidence: an observation (or series of observations) to be incorporated within his argument. These signs (observations) provide evidence, not necessarily conclusive, which supports his explanation of meteorological phenomena.

Aristotle's use of the term 'sign' may, on occasion, have a technical meaning somewhat different from ordinary usage (as when it is found in the phrase 'weather signs', where 'signs' provide indications of future atmospheric conditions). But it is significant that the 'signs' used in his meteorological explanations are, in principle, accessible to all observers. Aristotle uses the predictive value and success of weather signs as evidence that his meteorological theory is correct; his argument can be restated as: 'that haloes are a sign of rain is evidence that my account of rain is correct'.[116]

Analogy

That certain features of meteorological theory are readily comprehensible is reinforced by Aristotle's frequent deployment of analogies with common experience; such analogies attempt to offer empirical support for explanations of distant phenomena, by making comparisons to the more familiar. Different types of analogy are used in the *Meteorology*, as well as in other works in the Aristotelian corpus, notably the writings concerned with living beings.[117] In the *History of Animals* (486b17–21), he points to analogies between hands and claws, and between hair, feathers and fish-scales. These analogies suggest a similarity of function; this type of analogy is not unlike the modern notion of homology used to relate the parts of different species.

In the *Meteorology*, Aristotle draws different types of comparisons, pointing to similarities between those phenomena which are more familiar and those which are difficult to access. Some of his comparisons are rather tentative, perhaps only intended to be suggestive. Sometimes he takes it for granted that we supply the details of the analogy ourselves, presumably because he considers the event to be so common. For example, as part of his argument that the motion of the sun inflames air, Aristotle states that objects in motion are often found to melt, though he does not give specific examples. Other comparisons suggest close analogies, pointing to a strong resemblance between the phenomena being compared.

One of the deeper implications of the *Meteorology* is the extent to which the cosmos as a whole may be regarded as being like a living being. It is important to stress that while Aristotle repeatedly points to such analogies to living bodies, he does not seem to think of the cosmos itself as an animal, or living being.[118] Nonetheless, he deliberately draws an elaborate and extended analogy between exhalations trapped in the earth and wind trapped in the human body. To begin with, both can cause movements: 'we must suppose that the wind in the earth has effects similar to those of the wind in our bodies whose force when it is pent up inside us can cause tremors and throbbings, some earthquakes being like a tremor, some like a throbbing'.[119] Just as the causes of earthquakes can be understood as analogous to bodily processes, so can the saltiness of the sea, which is due to the dry exhalation. In living bodies, the residues produced are salty and bitter; urine and sweat are the examples given. Similarly, the dry exhalation is, he explains, a residue of natural growth and generation, and so is salty. The dry (salty) exhalation is mixed with the moist and vaporous exhalation, condenses into clouds and falls as rain. In this way, the sea contains salt, as a residue from the dry exhalation.[120]

The idea that the dry exhalation contains residues from generation and growth reinforces analogies drawn in several of his writings, between the earth and the means of nourishment in plants and animals. At several points in the *Meteorology*, Aristotle refers to the internal heat of the earth. For example, he states that 'there is in the earth a large amount of fire and heat' (360a6) and also mentions its 'internal heat' (360b32) or 'internal fire' (365b26). While there are several possible ways of understanding this internal heat,[121] Aristotle may have understood the earth as possessing some inherent heat. In relation to Aristotle's meteorological views, the sun (that is, its motion) plays an important role in heating the earth, whether or not the sun is the sole source of the earth's internal heat.

Aristotle explained that the exhalations arise when the earth is heated by the sun; a more vaporous exhalation comes from the moisture in and on the earth, and a dry exhalation from the earth itself.[122] The dry exhalation is hot and easily inflammable. Above the surface of the earth it produces winds, comets, thunder and lightning. Its motion under the surface of the earth gives rise to geological phenomena like earthquakes. Aristotle also explains that 'the sun not only draws up the moisture on the earth's surface, but also heats and so dries the earth itself'.[123] The dry exhalation apparently also heats the interior of the earth.[124] Aristotle may have thought of the hot exhalation circulating in the bowels of the earth as supplying the heat which allows the earth to function as a surrogate stomach.

The analogy between the earth and the stomach is explicitly formulated. In the *Parts of Animals*, Aristotle explains that 'plants get their food from the earth by means of their roots; and this food is already elaborated when taken in, which is the reason why plants produce no excrement, the earth and its heat serving them in the place of a stomach'.[125] He suggests that the stomach of animals is an 'internal substitute for the earth'.[126] Exhalations and heat arise when food is being digested,[127] just as exhalations and heat come from the earth when it is heated by the warmth of the sun.

In *On Youth, Old Age, Life and Death, and Respiration*, Aristotle emphasizes that the retention of heat by animals and plants is crucial, because 'everything living has soul, and it, as we have said, cannot exist without the presence of natural heat'. In plants, the natural heat is sustained both through nourishment (from the earth) and through the surrounding air. He explains the effects of changes in air temperature on plants by drawing an analogy with the ingestion of food by humans, which cools their bodies:

> For the food has a cooling effect when it enters (as it does for men immediately after a meal), whereas abstinence from food produces heat and thirst. The air, if it be motionless, becomes hot, but by the entry of food a motion is set up which lasts until digestion is completed and so cools it. If the surrounding air is excessively cold owing to the time of year, there being severe frost, the force of the heat dwindles; but when there are hot spells and the moisture drawn from the ground cannot produce its cooling effect, the heat comes to an end by exhaustion. Trees suffering at such seasons are said to be blighted or star-stricken. Hence the practice of laying beneath the roots stones of certain species or water in pots, for the purpose of cooling the roots of the plants.[128]

These analogies between the earth and various types of living things, especially in relation to nourishment and digestion, are powerfully drawn and evocative. They serve to emphasize the links between meteorological phenomena (including earthquakes and winds) and living things, and play an important role in underpinning Aristotle's meteorological views.

The analogy between the earth and digestion is only one of many employed in the *Meteorology*. So, for example, Aristotle draws an analogy between a shooting star and a spark igniting scattered chaff, to explain the different appearances of shooting stars and comets:

> the course of a shooting star is similar in that because the fuel is suitable it runs quickly along it. But if the fire were not to run through the fuel and burn itself out, but were to stand still at a point where the fuel-supply was densest, then this point at which the fire stops would be the beginning of the orbit of a comet. So we may define a comet as a shooting star that contains its beginning and end in itself.[129]

Tapping into everyday experience, Aristotle (369a20–5) uses another analogy to explain why thunderbolts, hurricanes and similar phenomena are seen to move downwards, in a direction opposite to what might be expected:

> for although all heat naturally rises, they must be projected away from the dense formation. Analogously, when we make fruit stones jump from between our fingers, they often move upwards in spite of their weight.[130]

He offers a similar homely analogy (369a30) to explain the noise of thunder which is produced when the windy exhalation in the clouds strikes a dense cloud formation, as being very like the familiar noise of a crackling flame, heard when the exhalation hits a log fire.

But, in spite of his frequent use of analogies, Aristotle (perhaps surprisingly) does not examine their logical character.[131] He did not offer an analysis of the use of analogy; he seems simply to take for granted the usefulness of analogies in helping to locate causes. At one point in the *Metaphysics*, he even suggests that in some cases an analogy will offer our best, and only, way of understanding: 'we must not seek a definition of everything but be content to grasp the analogy'.[132] Aristotle employs analogies from everyday experience not as part of a demonstration or proof, but rather to make the explanation comprehensible.[133] The references to the body and bodily processes, notably digestion, and comparisons such as the salty residue produced in dry exhalation with sweat and urine, would have provided homely and familiar examples.[134] The use of everyday examples and analogies provided an empirical basis for Aristotle's explanations of meteorological phenomena which are too distant or difficult to be investigated directly. The use of analogies to living things plays an important role in the *Meteorology*, particularly in his discussion of the exhalations central to his meteorological theories; these analogies link meteorology quite literally to the earth and help, for example, to explain seismological activity.

For many of the ancient authors writing on meteorology, the use of analogy is part of the attempt to provide empirical support for a hypothesis; this use is characteristic of ancient meteorological explanations, where direct investigation was not always possible. Throughout the *Meteorology*, Aristotle incorporated and engaged with various forms of indirect information, including observations reported by others (in some cases, traditional 'weather signs'), *endoxa* (reputable opinions) and analogies. But the use of this sort of information is by no means confined to the philosopher, or to scientific explanation. While everyday sorts of information (including received wisdom, in the form of *endoxa*, 'signs' and appeal to common experience) play significant roles in Aristotle's explanation of meteorological phenomena, in the *Meteorology* he also employed more unusual explanatory strategies, not normally found in everyday discourse. His use of 'experiments', diagrams and mathematical language are examples.

'Experiments'

In offering explanations of meteorological phenomena, analogies drawn from everyday experiences, as well as from more unusual situations, seem to have been adopted as the 'next best' approach, where no other empirical method of gaining information was possible. In some cases, the comparison is drawn from an experience or experiment constructed as a practical illustration of Aristotle's explanation, chosen to offer confirmation of a hypothesis. In considering Aristotle's use of evidence based on experiment, it is important to recognize that he was not concerned with modern issues such as the employment of controlled conditions, or the importance of repeatability and replication of results.[135]

Aristotle explains the saltiness of the sea as being due to the admixture of the dry and moist exhalations. The dry exhalation contains residues produced during natural processes of growth and generation; seawater, as an admixture, contains such earthy residues. As part of his explanation of seawater as an admixture, Aristotle refers to several experiences, which might be termed 'experiments'. In one, Aristotle states that, if a person were to mould a wax jar and place it in the sea, having fastened the mouth of the jar so that no water could get in, the water that did seep through the wax walls would be fresh, for the earthy substances causing the saltiness would have been filtered out. The substances that are mixed into seawater contribute to its weight and density; it is the density of the sea that allows ships to float which might sink in rivers. He then mentions that, if water is made very salty, by mixing in salt, then eggs will float on it; the sea contains a similar amount of material mixed into it.

Aristotle presents this information straightforwardly, suggesting that he has tried both procedures.[136] But in the first it appears that Aristotle has not actually carried out the experiment, for it doesn't work as reported. Wax is a usefully mouldable substance; as such, it is employed in casting, in the 'lost-wax' process. But wax is also non-porous; for this reason it is traditionally used as a water-proofing agent.[137] Aristotle repeated the claim in the *History of Animals* (8.2, 590a22–7) and it was also taken up by both Pliny (*NH* 31.37.70) and Aelian (9.64).[138] Modern scholars have often been critical of Aristotle's reported experiments in the *Meteorology*, and have suggested that he accepted the testimony of others, presenting the results as if he had tried the procedures himself. G.E.R. Lloyd has defended Aristotle, suggesting that, for example, it is possible that water vapour condensed in the wax jar once it was submerged.[139] While Aristotle may have done the experiment as described, it seems possible that he is reporting hearsay evidence, which literally does not hold water. In the second case, Aristotle's report is correct. Eggs will float on salty water. In the *Meteorology*, Aristotle uses everyday examples and analogies to explain phenomena. The egg experiment could be attempted by any interested reader or auditor at his lecture, and so reinforces the apparent familiarity of the comparison.

Diagrams

The view that the *Meteorology* was not a polished piece of writing, but is derived from lecture notes, is reinforced by the references to visual aids, including illustrations or diagrams presumably displayed to the audience, that occur in several places.[140] In his discussion of the Milky Way, Aristotle (346a32) refers to a *hupographē*, which illustrates the stars: 'The circle and the stars in it may be seen in the *hupographē*.' While the character of the *hupographē* is not certain, the reference must be to an illustration of some sort, presumably a drawing or a diagram. But the *hupographē* does not appear to be a lettered mathematical diagram.[141] The same term is used to describe the 'wind-rose' (diagram depicting wind direction) whose construction

Aristotle details in Book 2. Aristotle was interested in the problems of graphical representation: this is made clear by his criticism of the way in which contemporaries drew and mapped the earth, a subject he touches on in his treatment of the winds (362b12–15).[142]

In his discussion of the winds he includes a very detailed description of a *hupographē* (363a26–364a4), which is meant to aid in following the exposition of the winds' positions and the description of those which can blow simultaneously, as well as their names and number (363a21ff.). The illustration is described in sufficient detail in the text to allow it to be reproduced. The description begins as follows:

> For the sake of clarity we have drawn the circle of the horizon; that is why our figure is round. And it must be supposed to represent the section of the earth's surface in which we live; for the other section could be divided in a similar way.[143]

His opening words indicate the function of the diagram: it is meant to illustrate and augment the verbal description.

Characteristically, Aristotle sets out his points of reference: 'Let us first define things as spatially opposite when they are farthest removed from each other in space (just as things formally opposite are things farthest removed from each other in form); things are farthest removed from each other in space when they lie at opposite ends of the same diameter.' The description of the diagram has a mathematical flavour, as in this extract:

> Let the point A be the equinoctial sunset, and the point B its opposite, the equinoctial sunrise. Let another diameter cut this at right angles, and let the point H on this be the north and its diametrical opposite Θ be the south. Let the point Z be the summer sunrise, the point E the summer sunset, the point Δ the winter sunrise, the point Γ the winter sunset. And from Z let the diameter be drawn to Γ, from Δ to E.[144]

He employs geometrical terms (e.g. 'diameter', 'point', 'angle'), and the diagram is lettered.

But in spite of the seemingly mathematical features of the text and the diagram, the description is not purely mathematical, that is, the text is not about mathematical objects. Rather, it is meant to define the positions of the winds, as Aristotle explains: 'Since, then, things

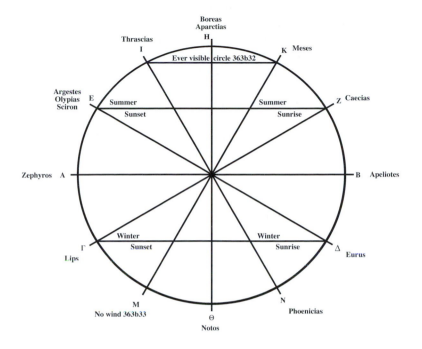

Figure 3.2 Modern diagram based on Aristotle's description of the winds.

Source: *Meteorologica*, trans. H.D.P. Lee (1952) Cambridge, Mass.: Harvard University Press, *Loeb Classical Library*, 187.

spatially farthest removed from each other are spatially opposite, and things diametrically opposed are farthest removed, those winds must be mutually opposite which are opposed diametrically.' He goes on to note that 'the names of the winds corresponding to these positions are as follows: Zephyros blows from A, for this is the equinoctial sunset. Its opposite is Apeliotes which blows from B, the equinoctial sunrise. Boreas or Aparctias blows from H, the north. Its opposite is Notos which blows from Θ, the south, Θ and H being diametrically opposed.' Aristotle goes through the rest of the winds, explaining their position on the diagram and their direction.[145] He does not comment on the nature of the diagram, except to underline its usefulness.

The positions of the winds are determined by reference to solar events, for example equinoctial sunrise and summer sunset. In his discussion of the winds, Aristotle explains (361a30–361b1) that 'wind is a body of dry exhalation moving about the earth'. He stipulates the material and efficient causes: 'though their motion takes its origin from above, the material from which they are produced comes from below'. The exhalation is produced from the earth, while the celestial motions determine the direction of flow. The sun plays an important role, for it both prevents and encourages the rise of the winds (361b14–362a31). The sun hinders wind by scorching the earth with its heat, so drying the earth too quickly to allow the exhalation to gather in any quantity (361b15–23). An example of how the sun encourages wind is found in the melting of the snow in the polar regions, which results in the blowing of the Etesian winds (362a12–31). The seasonal proximity of the sun to the inhabited regions of the earth (e.g. 362a25–6), and its daily appearance and disappearance during the day and night, affect the production of exhalation and so the winds' rising.

Aristotle does not indicate whether the use of the diagram, or wind-rose, is his own contribution to the identification of the winds.[146] However, the desirability and usefulness of a simple diagram to indicate the positions of specific winds are obvious. Such wind-roses (or their descriptions) survive in later, Roman authors (notably Vitruvius and Pliny), and in Roman (Imperial) artefacts.[147]

Aristotle also employs lettered diagrams in his discussion of haloes and rainbows, but his treatment of these seems to be different from that of the winds. For one thing, he refers to the lettered diagram that accompanies his discussion of the rainbow as a *diagramma*, rather than a *hupographē*. This difference in terminology may not be particularly significant, or it may indicate a different conception of what is being done in the treatment of the rainbow. *Diagramma* appears to have a particularly mathematical meaning in Aristotle; he uses the term to mean 'a mathematical proposition'.[148]

Geometry plays a central role in Aristotle's explanation of haloes; and in his treatment of the rainbow we also find a detailed geometrical discussion. In some ways, the introduction of mathematical argumentation seems sudden and unexpected. Why did Aristotle introduce mathematics into the *Meteorology*? The answer is that he has to, given his view of the nature of the study of the rainbow, which he regards, as he explains in the *Posterior Analytics* (79a11f.), as part of the field of mathematical optics; there he states that optics is a branch of mathematics.[149] In his discussion of haloes, rainbows, mock suns and 'rods' in the *Meteorology*, he explains that these are all phenomena of reflection: 'they differ in the manner of the reflection

Figure 3.3 Marble wind-rose (anemoscope), second to third century CE, found between the Esquiline and the Colosseum in Rome. The central hole may have been intended for a pennant or flag, to indicate wind direction.

Source: Musei Vaticani, Vatican City.
Note: See Obrist (1997), Lais (1894).

and in the reflecting surface, and according as the reflection is to the sun or some other bright object'.[150]

At various points in the *Meteorology* he makes it clear that there has been a lack of agreement about particular phenomena and whether or not they are optical. So, for example, in his review of theories about the Milky Way and comets, Aristotle mentions the views of predecessors who considered them to be optical phenomena, due to reflection (a view he rejects). Aristotle reports that Hippocrates of Chios and his student Aeschylus thought that the tails of comets were due to reflection, and were not part of the comet itself (342b36–343a20). The followers of Anaxagoras and Democritus maintained that the Milky Way is due to the light of certain stars (345a25ff.), while unnamed others regard the Milky Way as a reflection (345b10ff.). Aristotle argues against these theories; in his view, comets and the Milky Way are similar phenomena, composed of material of suitable consistency within the hot dry exhalation, in that part of the terrestrial region just below the celestial.[151] They are not optical phenomena, but are materially constituted. He explicitly contrasts comets with the haloes that appear around the sun and moon: 'the difference between them is that whereas the colour of the sun's halo is due to reflection, the colour of the comet's tail is what it actually appears to be' (344b7–8). The question of distinguishing between phenomena which are materially constituted and those which are optically produced by reflection comes up in the discussion, but is not the focus of Aristotle's treatment. So, while he states his view that rainbows, haloes, mock suns and rods are optical phenomena produced by reflection, he does not directly contrast these phenomena with those that have a material constitution. Nor does he organize his discussion around the distinction.[152]

As he begins his treatment of these haloes and rainbows, he indicates that the phenomena of reflection are properly studied as part of optics. Accordingly, these optical phenomena of reflection are subject to the sort of explanation provided by the branch of mathematics known as optics. In the *Posterior Analytics* (79a10–12), Aristotle notes that 'related to optics as this is related to geometry, there is another science related to it – viz. the study of the rainbow'.[153] In the *Metaphysics* (995a15–17), he explains that mathematics has a special role to play in the discussion of those things that are not materially constituted:

> The minute accuracy of mathematics is not to be demanded in all cases, but only in the case of things which have no

matter. Therefore its method is not that of natural science; for presumably all nature has matter.[154]

So, optical phenomena like haloes and rainbows, which are not materially constituted, are appropriately studied by mathematics.

But the relationship between mathematics and physics is not as easily understood as this passage might lead us to believe. Aristotle repeatedly grappled with the problem of defining the objects of study of mathematicians, and the boundaries between mathematics and physics.[155] As Edward Hussey has emphasized, 'Aristotle was not the first to suggest that the physical world exhibits mathematical relationships of various kinds.'[156] It is clear from Aristotle's discussions that, by the time of his writing, several branches of mathematics had been developed with applications to the physical world, including geometry, optics, harmonics and astronomy. At various points, he discusses the nature of mathematics and the objects studied by mathematicians. In some cases, for example the study of celestial bodies, Aristotle's view seems to be that physics and mathematics deal with the same (sensible) things, from different perspectives. Mathematics is an appropriate, even necessary, means through which to study some aspects of the physical world.

In the *Physics*, Aristotle considers 'how the mathematician differs from the physicist. Obviously physical bodies contain surfaces and volumes, lines and points, and these are the subject-matter of mathematics.'[157] He makes it clear that some types of mathematics are 'more physical' (τὰ φυσικώτερα τῶν μαθημάτων) than others; these include optics, harmonics and astronomy, which are all related to geometry.[158] The relationship between optics and geometry serves as an example: 'while geometry investigates natural lines but not *qua* natural, optics investigates mathematical lines, but *qua* natural, not *qua* mathematical'.[159] So, geometry treats lines, but not as the boundaries of physical objects; optics studies lines, as they relate to vision and rays of light.[160]

In the *Posterior Analytics*, Aristotle discusses the relationship between different areas of study. He states that 'it is for the empirical scientists to know the fact and for the mathematical to know the reason why; for the latter have the demonstrations of the explanations'. According to Aristotle, physics provides 'the fact that' (τὸ ὅτι) and mathematics offers 'the reason why' (τὸ διότι).[161] As part of this discussion, he specifically refers to the study of the rainbow, which he explains is related both to optics and to geometry. In his view, mathematics is necessary to the full explanation of the rainbow; physics alone cannot offer sufficient understanding.

And in his treatment of those meteorological phenomena which he specifically describes as due to reflection, namely, haloes and rainbows, Aristotle relies on geometry for his explanation. He addresses the question of the shape of the halo, explaining 'why it is round and why it appears round the sun or moon or similarly round one of the other stars', by discussing the reflection which occurs when air and vapour are condensed into cloud.[162] Reflection, an optical phenomenon, is fundamental to the explanation: 'our vision is reflected from the mist which condenses round the sun and moon; which is why a halo does not appear opposite the sun like a rainbow. And as the reflection is symmetrical on all sides, the result is bound to be a circle or a segment of a circle.' Note that his description relies on the geometrical language of circles and symmetry. Aristotle then launches into a detailed geometrical discussion:

> let the line AΓB, AZB, and AΔB be drawn from the point A to the point B, each forming an angle; let the lines AΓ, AZ, AΔ be equal to each other, and the lines drawn to B, that is ΓB, ZB, ΔB, also equal to each other. Let the line AEB be drawn and the triangles so formed will be equal as they stand on the equal base AEB. Let perpendiculars be dropped from the angles to AEB, ΓE from Γ, ZE from Z, ΔE from Δ. These perpendiculars are then equal, being in equal triangles and in one plane. For all meet AEB at right angles and at the single point E. The figure thus drawn will be a circle with centre E.[163]

In the *Posterior Analytics*, Aristotle explains that 'mathematics is about forms; for its objects are not said of any underlying subject – for even if geometrical objects are said of some underlying subject, still it is not *as* being said of an underlying subject that they are studied' (79a8f.). But even though the treatment is geometrical, the study of these phenomena is a type of mathematics which is 'more physical'.[164] In his explanation of the halo, he makes it clear that geometry is being used to describe and explain natural phenomena: 'B is of course the sun, A the eye, and the circumference drawn through ΓZΔ the cloud from which the vision is reflected to the sun'.[165] This treatment of haloes employs geometrical language and figures but, strikingly, the geometry is very deliberately not only about figures, shapes and solids; rather, it is explicitly linked to the physical bodies

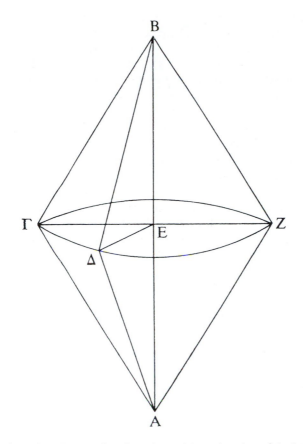

Figure 3.4 Modern diagram based on Aristotle's explanation of the halo.

Source: *Meteorologica*, trans. H.D.P. Lee (1952) Cambridge, Mass.: Harvard University Press, *Loeb Classical Library*, facing 249.

(the sun and the eye) involved in the production and perception of optical phenomena. That Aristotle's lengthy treatment of the rainbow (at 373a32–377a28) is similar in style to his treatment of haloes is not surprising, since in his view the rainbow is also a reflection and is therefore properly studied by mathematical methods appropriate to optics.[166] In his detailed discussion, geometry is central to the explanation of various features of the appearance of rainbows, including their coloured bands (375b9–11) and semicircular shape (375b16–377a28). There are also detailed references to diagrams (which do not survive).

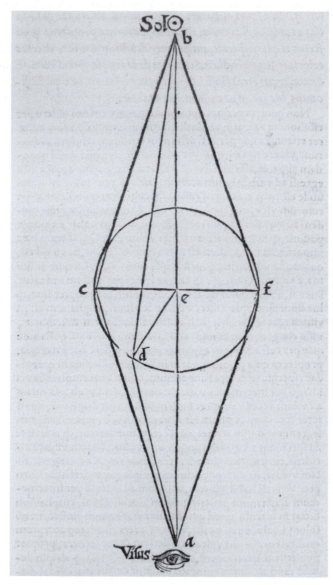

Figure 3.5 Diagram accompanying an edition by Alessandro Piccolomini of Alexander of Aphrodisias' commentary on Aristotle's *Meteorology* (Venice: H. Scotus, 1548); by permission of the Syndics of Cambridge University Library.

Source: Cambridge University Library.
Note: The location of the sun and the eye are labelled.

Reviel Netz, as part of a broader consideration of the structure of mathematical proofs, has noted that the passage at 373a is especially dependent on a diagram, with not a single letter fully defined; the diagram may have been used in a lecture.[167] Netz noted that there are no obvious precedents for Aristotle's practice, and that his use of the diagram in this way was an isolated phenomenon, even within the school he founded.[168] But we should not be surprised that Aristotle's writings provide the earliest example of the use of a lettered diagram. After all, he sought to bring coherence to the study of nature (as well as other areas) and, as part of his overall programme, strove to provide definitions of technical terms and methods;[169] he prided himself on carrying the field forward.

Remember that Aristotle's treatment of optical phenomena is set within a broader discussion of the exhalations; as *meteōra*, it makes sense that the rainbow and halo are treated in the *Meteorology*. Netz points out that it is unusual (but not unknown) to find a mathematical text embedded in a discursive setting, but he regards the passage referring to the rainbow as 'hardly worthy of being called mathematics'.[170] Analysing the structure of the proof, he points out that Aristotle 'does not distinguish firmly between construction and proof'; a geometrical proof must be clearly separated from the construction, in order to be developed in an orderly manner. Furthermore, it is necessary for the diagram and text to be set apart. Netz highlights what he regards as the most striking feature of Aristotle's treatment:

> one thing that is clear about this proof is that it is not compartmentalised from a more general, non-mathematical discussion. The proof starts immediately from a discussion of the rainbow, and ends, just as immediately, with an identification of the letters in the proof with the topics under discussion – eye, sun, cloud. The proof is embedded within a larger discursive context, and borrows the discursiveness of this context.[171]

This is precisely what is so interesting about Aristotle's use of diagrams within the explanations of the halo and rainbow: that they are not separated from the body of the discussion, but are included as a necessary part of the explanation. While Aristotle's presentation may not meet the standards of some mathematicians, he is clear that these non-materially based optical *meteōra* must be explained using optics and geometry. The use of diagrams as a crucial part of the explanation of haloes and rainbows may reinforce our understanding of the *Meteorology* as a series of lectures, for which the diagrams referred to served as visual aids.[172]

It is important to keep in mind that the mathematical treatment of haloes and rainbows in Book 3 is set within the broader explanation of the effects of the exhalations. Aristotle himself reminds us of this (378a12ff.) when he concludes his discussion of other phenomena that are also due to reflection – 'rods' ('sun pillars') and 'mock suns' (parhelia) – by announcing that 'this completes our enumeration of the kind of effects produced by exhalation in the regions above the earth's surface'.

He then turns to a brief discussion of things produced by exhalation below the surface of the earth. Just as there are two exhalations which produce various meteorological phenomena above the surface of the earth, so too the two exhalations produce two different kinds of subterranean material: 'there are, we maintain, two exhalations, one vaporous and one smoky; and there are two corresponding kinds of body produced within the earth, "things which are dug up" [ὀρυκτὰ] and metals'.[173]

Aristotle very briefly discusses the products of the two exhalations formed within the earth; the dry exhalation produces those stones which are dug out and cannot be melted, such as cinnabar, while the vaporous exhalation produces metals, such as iron. Underground products of exhalations undergo similar processes of production to those formed above the earth; Aristotle draws an analogy between the compression and solidification of metals and the formation of dew and frost, through condensation and freezing. Following his comments on the subterranean products of the exhalations, Aristotle concludes Book 3 of the *Meteorology*.[174]

Throughout the work, Aristotle presents an account of meteorological phenomena which is based on his theory of dual exhalations; his treatment includes a very broad range of phenomena, from those which occur very high up above the earth (such as comets) and others which are subterranean. As he states at the beginning, he is concerned with 'all phenomena that may be regarded as common to air and water, and the various kinds and parts of the earth and their characteristics'.[175] He regards his treatment of meteorological phenomena as following on (conceptually) from his work on the first causes of nature, natural motion, celestial motion, the elements and generation and destruction in general; and it precedes his treatment of animals and plants. His explanation of the causes of meteorological phenomena is clearly part of his larger philosophical programme; in order to fulfil his own philosophical aims, he must give an account of this broad group of phenomena, which he regards as occurring naturally, but with less regularity than those in the celestial region. To us, the range of phenomena explained in the *Meteorology* may seem to be unrelated. But, for Aristotle, these phenomena are to be understood as being due to the two exhalations, which can be understood as providing a single material cause and a single explanatory principle.[176]

At the opening of the *Metaphysics* (1.1), Aristotle states his view that 'all men, by nature, desire to know'. Reading through the *Meteorology*, the reader is left with the sense that the phenomena *are* broadly comprehensible, in spite of the warning at the beginning of the work that we will not be able to fully understand all of them. As part of his explication of the phenomena, Aristotle employs familiar tactics. For example, the consideration of *endoxa* and the use of analogies with everyday experience are commonly used tactics, not restricted to natural philosophers; many people cite reputable opinions and draw homely analogies when they offer explanations. But Aristotle did not shy away from more specialist approaches, including the use of experiments, geometrical language and diagrams in his explanation of the *meteōra*. Despite the difficulty of explaining meteorological phenomena, through his use of a variety of tactics, Aristotle offers a relatively simple account of a broad range of phenomena, from lofty comets to subterranean ores, based on the effects of the two exhalations. He emphasizes the tentativeness of his explanation, but justifies its acceptance, as a rational and possible account:

> We consider that we have given a sufficiently rational explanation of things inaccessible to observation by our senses if we have produced a theory that is possible: and the following seems, on the evidence available, to be the explanation of the phenomena now under consideration.[177]

The *meteōra* may not be absolutely comprehensible, even to natural philosophers, but they can still be rationally and plausibly explained. In the *Meteorology*, Aristotle indicates that he is content to offer just such a rational account.

Theophrastus

Aristotle's colleague and successor as head of the Lyceum, Theophrastus of Eresus (372/1 or 371/70–288/7 or 287/6 BCE) is associated with several short works on meteorological topics. Like Aristotle, Theophrastus' work ranges over a broad spectrum of topics, but, with the exception of his work on botany, relatively little survives. The authorship of *On Weather Signs* is not firmly established, and it is possible that neither of his other works, the *Meteorology* and *On Winds*, is complete.[178] But there is evidence that Theophrastus' approach to meteorology was influential; this will be examined in the following chapter.

Theophrastus' *Meteorology* is no longer extant in Greek. This work is also known as the *Metarsiology* (= *Metarsiologica*), a variant form which refers to 'things in the sky'.[179] One Syriac translation (only preserved in a fragment) and two Arabic translations survive, of which the fuller and more recently discovered one, apparently by Ibn al-Khammar (b. 331/942 or 943, d. 421/1030), is based on the Syriac. There has been a good deal of debate as to whether the Syro-Arabic *Meteorology* is incomplete, or based on an excerpt.[180] The work is somewhat patchy, and few connections are made between the various subjects discussed; further, there seems to be no logic in the order in which the material is presented.

Scholars are divided about the original form of the work. Some have regarded the *Meteorology/Metarsiology*[181] as a sort of doxographical treatise, in which Theophrastus presented the views of other thinkers, as well as his own. Hans Daiber argues that what survives in translation is an accurate rendering based on a Greek original, which was a draft or sketch for a lecture, in which Theophrastus intended to present various ideas by means of examples. Jaap Mansfeld has argued that the text published by Daiber is actually an abridgement of the *Metarsiology*. David Sedley has suggested that Theophrastus may have presented roughly the same material on meteorology in two different works, the two-book *Metarsiology* (*Metarsiologica*) and the now-lost doxographical *Physical Opinions*.[182] In other words, there is no real agreement as to the extent to which what we have of Theophrastus' *Meteorology/Metarsiology* represents his full treatment of the subject.

Within the surviving text, the discussion of phenomena is divided into those which occur above the earth (including thunder, lightning, thunderbolts, clouds, rain, snow, hail, dew, hoar frost, winds and haloes) and those below (that is, earthquakes). This simple division may have been most convenient for a lecture or oral presentation.[183] Theophrastus discusses the causes of the phenomena in the following order: thunder, lightning, thunder occurring without lightning, lightning without thunder, the reasons why lightning precedes thunder, thunderbolts, clouds, different kinds of rain, snow, hail, dew, hoar frost, different winds, the halo around the moon and earthquakes.[184] The *Meteorology/Metarsiology* is almost presented as a list of phenomena and their causes; this list-like format also suggests that the text preserves notes for a lecture (or lectures).

The list has several features worth noting: in most cases, a series of causes is given for each phenomenon and, in many cases, a direct appeal is made, via analogy, to everyday experience. These two explanatory strategies, the presentation of multiple causes and the analogy with everyday experience, characterize Theophrastus' approach.

For example, he begins with an account of the causes of thunder; he gives seven: (1) when two hollow clouds collide, (2) when wind enters a hollow cloud and turns it, (3) when fire falls into a humid cloud and is extinguished, (4) when wind strikes a broad and icy cloud violently, (5) when wind enters a long, crooked, hollow cloud, (6) when wind is congested in a hollow cloud and the cloud splits open and (7) when rough clouds rub together. Four causes are given for lightning, three for thunder occurring without lightning, two for lightning without thunder, and two reasons why lightning precedes thunder; four causes are given for earthquakes.[185]

Ancient authors had a variety of reasons for listing and discussing a number of possible causes of phenomena.[186] Aristotle sometimes described the ideas of his predecessors, often in order to argue against them. Other authors, including Theophrastus himself, collected the opinions (δόξαι) of others for academic purposes; collecting and listing and properly attributing the ideas of others was part of the doxographical tradition of scholarship.[187] However, Theophrastus, in the *Meteorology/Metarsiology*, does not attribute the various causes to other thinkers by name, and it does not seem appropriate to regard the listing of causes as part of his doxographical work.[188] The assessment of Robert Sharples seems reasonable, that 'Theophrastus' practice goes beyond, though it can be seen as a development from, the collecting of different explanations for purposes of debate or discussion, whether for its own sake or with a view – as often in Aristotle – to selecting one from among many explanations or to replacing them by a different explanation altogether.'[189] Theophrastus seems to have held the view that, for some phenomena, a variety of causes were at work,[190] and to have been interested in correlating 'different explanations with different forms of a phenomenon in our experience'.[191]

However, not all the phenomena are explained by a multiplicity of causes, nor are all the explanations given in a list. So, for example, only one explanation is provided for hail, which occurs when 'drops of water are transformed and hardened by coldness'. The discussion of rain is strikingly brief: 'Heavy rain occurs, if very hard winds squeeze and accumulate the clouds. Continuous rain occurs, if many vapours ascend from the seas.'[192]

This brevity is in marked contrast to the lengthy explanation of thunderbolts. First, the thunderbolt is described as being either a windy fire or a fiery wind. Next, a further description of thunderbolts and their effects is given, with care taken to contrast thunderbolts with other sorts of fire. Then, two causes are given for thunderbolts: fire being hidden in a cloud and suddenly slipping away, and wind being hidden in a cloud and catching fire as it circulates. Two causes account for thunderbolts reaching us: the bottom of a cloud being split, or storms or winds beating on top of a cloud. Theophrastus goes on to elaborate why it is that cloud is split from the bottom rather than the top, why thunderbolts are more frequent in spring, and why they are more frequent in high places. The section ends with an explanation of why it is that, when a thunderbolt falls on a purse holding coins, the purse itself is not affected but the coins melt.

The discussion of thunderbolts contains many references to everyday experiences, with analogies from ordinary life. So, for example, he explains that when thunderbolts come together they form a single thunderbolt, in the way that a single river is formed from many fountainheads, 'when the water coming from them is assembled in one place'.[193] In his descriptions of the ways in which clouds may be cut or split, or extended by wind in their interior, he repeatedly cites experiences with skins or bladders, being burst when squeezed or strongly inflated, becoming full of cracks, or being irregular and splitting in the weaker sections.[194]

The section on thunderbolts is, to some extent, organized as a set of responses to questions. He suggests that the question about thunderbolts striking purses is a query that might be raised: such a question might be posed during a lecture. And the organization of this section is reminiscent of the questions and answers presented in 'problem' texts (*problēmata*), with the difference that Theophrastus' answers are presented as solutions that can operate simultaneously, while in the 'problem' texts some of the 'answers' given may be incompatible.[195] At several points throughout the text, Theophrastus indicates that he is providing answers to possible questioners or even sceptics. In the first section, on thunder, he gives a summary conclusion: 'thunder arises because of the causes that we have mentioned'. He then turns to a possible question, saying 'suppose someone is sceptical and asks: "How is it possible that noise arises from clouds since they are not solid like stones and earthenware but rarefied like wool, whereas noise cannot arise from wool? For noise does not arise if a man beats tufts of wool, one with the other".' The imagined sceptic raises objections based on everyday experience; Theophrastus is prepared to answer this objection in the same terms:

> We can answer: We too do not maintain that noise arises in clouds because they are solid and similar to stones. We say however: even if the clouds are rarefied and split they are still able to make much noise. Similarly we can see among things that are visible solid things that produce no noise, like a pot of clay and lead. For these things do not have any causes which result in the making of much noise. And (we can see) among (visible) things rarefied things that produce much noise, like water and dry leaves. – Because wool is rarefied it does not produce any noise, not by virtue of being rarefied but because the causes which produce noise do not exist in it. This is our answer to that sceptical question.[196]

Elsewhere in the *Meteorology/Metarsiology*, he also poses hypothetical objections and provides possible answers. In the section on thunderbolts, he provides answers to someone wanting to know why thunderbolts are more frequent in spring and more frequent in high places; the discussion about a thunderbolt hitting a purse is also an answer to a potential question. In the final section, on earthquakes, Theophrastus gives a response to someone demanding 'the reason why some earthquakes are a kind of trembling, some a kind of inclination with an inclining movement and some only a kind of inclination (without an inclining movement)'.[197] In each case, the possible objection appears to arise from observed experience, rather than from theoretical concerns; the answers are framed in the form of analogies with everyday life.

The use of analogies relating the explanation of meteorological phenomena to everyday experience occurs throughout the text, beginning in the first section, on the seven causes of thunder. Here, as each cause is described, Theophrastus draws an everyday analogy. The first cause of thunder occurs when two hollow clouds collide and one strikes the other. Theophrastus states that 'we can observe something similar', for 'when we make our hand hollow and strike the one against the other, a great noise is the result'. Similarly, for the remaining causes of thunder, he points to common experiences of loud noises: wind blowing and entering caves and large jars, an ironsmith throwing glowing iron into water, wind striking paper, butchers blowing up guts, a bladder filled with air and then punctured, millstones rubbing against each other. So, too, in the listing of the four causes of lightning, each cause is followed by a direct analogy with everyday experience. The first two causes listed, beating and friction, are compared to the actions of stones being beaten against each other and sticks being rubbed together, which both produce fire. The extinguishing of fire in a humid cloud is like the action of an ironsmith plunging red-hot iron into water. Finally, the 'squeezing out' of fire from a cloud is like a sponge being squeezed, but the result of the first is fire, not water. Theophrastus' explanation of thunder contrasts with that of Aristotle, who regards thunder and lightning as the products of exhalation (Aristotle *Meteorology* 369a10–369b12).[198]

While Theophrastus often uses analogies with everyday experience within the *Meteorology/Metarsiology*, he also refers to direct observation. The formation of snow is explained not by analogy but by appeal to experience: Theophrastus points out that we can easily see that snow contains a great deal of air. Snow is formed when the cold freezes clouds before they turn into water, and before one part of the water is connected with the other; the water must be dispersed into very small drops which are separated by air. Theophrastus says that 'we can see with our own eyes' that snow contains much air. He offers as evidence the experiential 'proof' of compacting soft snow in the hand; as it is compacted, the quantity becomes less, and when it melts only a (relatively) small amount of water is left. The whiteness of snow is explained by analogy to other white bodies, such as foam, containing much air.[199]

Theophrastus discusses the causes of winds at some length; this is the second longest section of the work, after that on thunderbolts. He explains that 'wind is formed from vapor which is composed of fine and thick (parts)'. 'Winds arise from above and from below; their generation from below is either from water or from earth.'[200] He goes on to describe the way in which wind moves, and suggests a way (a sort of 'experiment') to help understand and illustrate such movement: 'when we place a tube on the surface of water and extract the air in it by sucking it with the mouth and by using the force of the vacuum, the water ascends and fills the tube, so that in it no vacuum is left. In this manner the winds are generated from below.'[201] Details of various winds (strong, continuous, hot and cold, dry and humid) are briefly reviewed. Theophrastus ends the section on the causes of winds by considering those that affect ships. He answers a hypothetical questioner asking about 'the wind which is called "WRS"', (presumably the Arabic transliteration of *euros* (εὖρος)), which pushes ships forward violently. Once again, the information is conveyed as a response to a query. He then offers a report about the *prēstēr*, and its potential effects on ships.[202] The practical concerns of seafarers and navigators are answered as part of his more general explanation of the causes of winds.[203]

Theophrastus also discussed the winds in a separate treatise devoted to the subject. His *On Winds* begins with a reference to a previously given explanation of their origin. He states that 'we have earlier considered the nature of the winds: of what they consist, in what way they come to be, and by what they are caused'.[204] The reference appears to be to the explanations offered in the *Meteorology/ Metarsiology*, and supports the view that *On Winds* was written after the *Meteorology/Metarsiology*.[205]

In *On Winds* Theophrastus saw as part of his task the elaboration of the characteristics and effects of particular winds, the associated or accompanying phenomena, and differences between particular winds. For Theophrastus, the study was fundamental, for 'what happens in the sky, in the air, on earth and on the sea is due to the wind'.[206] Theophrastus regards winds as the movement of air. In his view, wind moves in order to restore the air's balance, disrupted by the influence of the sun.[207] The exhalation (ἡ ἀναθυμίασις) is crucial to the generation of wind, but the sun plays a role as well; its rising and setting seems 'both to set the winds in motion and to halt them'.[208] The moon acts as a weak sun, also affecting the winds. Because of this, winds are stronger during the night and at the time of the full moon. As part of his discussion of the effect of the sun and moon on winds, Theophrastus asks whether these things occur also in conjunction (κατὰ σύμπτωμα), like the phenomena at the risings and settings of stars; he suggests that the matter is worthy of further inquiry.[209] In *On Winds*, he regards the winds as regular, predictable phenomena, even though not all details have been worked out to explain them. As elsewhere, Theophrastus uses everyday analogies as part of his method of explanation in *On Winds*. For example (at *On Winds* 19–20), the difference between expelling air through an open or closed mouth is offered as an analogy to the way in which the motion of the sun can be implicated in the production of hot or cold winds.

In discussing the causes of meteorological phenomena, he makes it clear that he is not aiming at providing one explanation for each type of phenomenon. On the contrary, for most meteorological phenomena, several different causes are named. The emphasis on the possible variety of causes is characteristic of his approach and was influential for later writers.[210]

This emphasis on multiple causes is one of the important differences between Theophrastus' approach to explaining meteorological phenomena and that of Aristotle.[211] Theophrastus does not show Aristotle's concern with providing a single, correct explanation. Rather, through his presentation (and acceptance) of multiple possible causes, Theophrastus signals that he is not constrained to offer only one explanation for a meteorological phenomenon; he is content to recognize several different explanations.

The relationship of Theophrastus' work to that of Aristotle has been the subject of much debate. Sharples has argued that, whatever their differences on individual points of doctrine, Theophrastus 'shared in, continued, and extended' Aristotle's work in every area. Moreover, he carried on Aristotle's philosophy in a fundamentally philosophical manner, by continuing to ask questions and raise difficulties.[212]

For Theophrastus, 'it is the task of science to grasp what is the same in several things'. This grasping of 'the same' in different things can be accomplished because things can be the same in several different ways: either 'in essence or in number or in species or in genus or by analogy'. Recognizing the various methods of acquiring knowledge, Theophrastus specifies analogy as one of the means by which we comprehend, explaining that 'analogical identity spans the widest interval, as though there were here the greatest distance between the objects, this appearance of distance being due sometimes to ourselves, sometimes to the object, sometimes to both'.[213] He notes that the use of analogy is particularly helpful in gaining understanding of difficult things:

> if there are even certain things that are known by being unknown, as some maintain, the manner of inquiry into them would be one peculiar to them, but needs some care to distinguish it from others; though perhaps, in cases where it is possible, it is more appropriate to describe them by analogy, rather than by the very fact that they are unknown — as if one were to describe the invisible by the mere fact that it is invisible.[214]

But even the use of analogy will not provide complete knowledge. Theophrastus points to the limits of what can be known and derides those who desire the impossible: 'those who demand proof of everything destroy proof, and at the same time knowledge'. In his view, 'it is truer to say that they seek proof of things of which there is not and from the nature of the case cannot be proof'.[215] But though he denied the possibility of proof in some cases, he still advocated the pursuit of knowledge. This is demonstrated by his willingness to continue to raise questions, and to point to interesting problems for further study.[216]

In his *Metaphysics*, Theophrastus raises questions, but also lays out his own programme of inquiry. He explains that there are different kinds of knowledge: of primary things, of natural things, of animals, of plants and of inanimate things. Each type has its own appropriate methods.[217] Theophrastus contrasted the study of first principles with the study of nature, noting that the study of natural philosophy is more varied and less ordered than metaphysics, for it is concerned with all sorts of things, not only the certain and unchanging. For this reason, he suggests, some people regard metaphysics as a more dignified area of study than natural philosophy.

Theophrastus himself seems not concerned particularly with questions of dignity, for his extant writings include several on subjects which had, apparently, not attracted the attention of other philosophers. So, for example, his writings on botany, stones and fire focus on areas which had not been investigated by others, including Aristotle. John Vallance has suggested that Theophrastus, in taking up these areas of inquiry, showed an unusual interest in engaging with the 'more varied' and less orderly; this distinguishes his work from that of Aristotle.[218] And while there are similarities between their meteorological explanations, Theophrastus' work differs significantly from Aristotle's in content, form and method.

Theophrastus' *Meteorology/Metarsiology* is organized according to the four elements, Fire, Water, Air and Earth, while Aristotle's discussion is centred on his two exhalations. Theophrastus does, to some extent, incorporate Aristotle's exhalations into his meteorological explanations. Fine vapour forms winds, while clouds arise from moist vapour; earthquakes can be caused by winds in the bowels of the earth.[219] However, for Theophrastus, the exhalations typically provide only one of several possible causes for any given phenomenon. So, for example, according to Theophrastus, wind being shut up in the earth is only one possible cause of earthquakes; he lists three others which do not involve wind. In fact, the three other causes involve each of the three other elements: earth being dry and crumbling or humid and dissolving, water being shut up and moving and causing the earth to shake, and fire being shut up and looking for a larger place. So the four elements play an important role in Theophrastus' explanation.[220] His explanation of wind as moving air may at first seem simple, but he offers a rather lengthy study of winds. While Aristotle in the *Meteorology* (1.13) had rejected the concept of wind as moving air as too simplistic, he then offered a simple cause, namely, the rotation of the heavens (2.4, 361a22–b1). Although Theophrastus' explanation may at first seem simple, he then goes on to consider various effects on the movement of air, including the motions of the sun, moon and stars.[221] His call for further research reinforces the sense that he is interested in a detailed understanding of the phenomena, including different varieties of each. The multiple explanations he offers suggest that he has investigated a range of causes.

The use of multiple causes is a hallmark of Theophrastus' meteorology, but he also incorporates other explanatory methods and techniques, including a sort of 'experiment', in some cases to help reinforce an analogy. His resort to analogy with common experiences is reminiscent of the approaches of his predecessors, including Aristotle, and is found elsewhere in his work, particularly his botanical writings.[222] But just as he multiplies possible causes of meteorological phenomena, so does he provide more examples of analogical explanation; familiar analogies are each linked to a particular possible cause. The emphasis on multiple explanations may be related to Theophrastus' view that all four elements play important roles in the causation of meteorological phenomena. It may have been his view that, because there are multiple elements involved, there must be multiple causes. Theophrastus' willingness 'to admit to the variety of nature even at levels that do not submit to general explanation' is the key difference between his approach and that of Aristotle.[223] This willingness to live with uncertainty, variety and 'disorderliness' in his study of nature may be linked to his readiness to adopt a multiplicity of causes to explain meteorological phenomena. As will be seen in what follows, Theophrastus' approach to explaining meteorological phenomena was very influential.

4

METEOROLOGY AS
A MEANS TO AN END:
PHILOSOPHERS AND POETS

Following Aristotle and Theophrastus, later philosophers who sought
to explain meteorological phenomena worked within a framework
inherited from them, in which certain phenomena were recognized
as meteorological and various methods of investigation and explanation were deemed appropriate. But, while both Aristotle and
Theophrastus included their work on meteorology within a larger
programme of specifically natural philosophy, for later philosophical
schools, particularly the Epicureans and Stoics, the explanation of
meteorological phenomena was undertaken primarily as a means to a
broader philosophical end, and the attainment of an ethical aim.

Theophrastus' approach to explaining meteorological phenomena
was very influential; it was brought to a wider public, auditors and
readers, through the works of widely read ancient authors, including
the poet Lucretius (who probably encountered Theophrastus' ideas
through Epicurus).[1] No Greek version of Theophrastus' *Meteorology/
Metarsiology* survives. A Syriac translation survives only in a fragment;
the fullest text is an Arabic translation, based on the Syriac.

At the end of the section on the causes of the haloes around the
moon, there is a discussion (and rejection) of the concept that a god
causes and intervenes in meteorological phenomena. The author
directly confronts the question whether some deity may be the cause
of meteorological phenomena (in this case thunder):

> Neither the thunderbolt (pl.) nor anything that has been
> mentioned has its origin in God. For it is not correct (to say)
> that God should be the cause of disorder in the world; nay,
> (He is) the cause of its arrangement and order. And that is
> why we ascribe its arrangement and order to God {mighty
> and exalted is He!} and the disorder of the world to the
> nature of the world. And moreover: if thunderbolts originate

in God, why do they mostly occur during spring or in high places, but not during winter or summer or in low places? In addition: why do thunderbolts fall on uninhabited mountains, on seas, on trees and on irrational living beings? God is not angry with those! Further, more astonishing would be the fact that thunderbolts can strike the best people and those who fear God, but not those who act unjustly and propagate evil. It is thus not right to say <about> hurricanes that they come from God; (we may) only (say the following) about something that happens to us to our harm or that diminishes divine power: It happens without any order. Consequently there is no indication of passing away in the case of God and any indication of being like an angel (= god-like) is to be removed from us.[2]

The possibility that a god is the cause of weather phenomena is rejected. This rejection of a divine role in the causation of meteorological phenomena occurs in a section of the work which Daiber, the modern translator, describes as an 'excursus', which interrupts the flow of the text and may be a digression.[3] At first reading it is not immediately clear who is responsible for this rejection, whether it is Theophrastus himself, the medieval translator, or some ancient intermediary.

Nevertheless, there are clear links with earlier portions of the text; reference is made to questions for which Theophrastus has already provided answers, namely, why thunderbolts are more frequent in the spring and in high places. These suggest that either the same author (Theophrastus) is responsible for the excursus as well as the text, or that the author (presumably a later, ancient intermediary or the medieval translator) of this particular passage had read the rest of the text very closely. Here, as in other works by Theophrastus, there do seem to be discrepancies within the conception of a divine agency in regular occurrences in nature. But Theophrastus does reject the notion of a divine agent using meteorological phenomena as a means of punishing errant mortals.[4]

Many scholars have argued that Greek natural philosophy excluded the gods as agents in the natural world; but this is an over-simplification. Certainly, in most of the ancient philosophical schools there was an assumption of some 'divine' presence in the cosmos (in some cases, the cosmos itself). The nature of this 'divinity' (or 'divinities') was a concern for ancient philosophers. However, it is important to recognize that what is 'divine' was not necessarily understood as a personal god. Nevertheless, within some philosophical schools, the idea of the existence of gods was accepted; their role was a matter for debate. Questions about whether the gods play a role in causing meteorological phenomena recur in both the Greek and Latin traditions.[5] Several ancient authors rejected the idea that gods control these phenomena as a punishment for humans. One of the arguments supporting Theophrastus' authorship of the 'excursus' is the similarity of the ideas here to those in later ancient authors, who presumably were familiar with his work.

Epicurus and Lucretius

There is strong evidence that the Greek philosopher Epicurus (341–271 BCE) and his follower, the Roman poet Titus Lucretius Carus (probably 94–55 or 51 BCE?), were influenced by Theophrastus' approach in their own discussions of meteorology.[6] Epicurus was the founder of an important school of philosophy. The practical goal of his philosophy was the achievement of calm and freedom from anxiety (*ataraxia* = 'being undisturbed'). In order to help achieve *ataraxia*, Epicurus developed a materialist philosophy which explained the world; the basis of the material composition of the universe, according to Epicurus, is atoms and empty space (the void). Epicurus denied the possibility that the gods have any influence in the world. Epicurus wrote a great deal, but only a few short works survive. Three of these are didactic letters;[7] the fourth, the 'Key Doctrines', is a set of epistemological and ethical maxims. These works survive because they were quoted, in their entirety, in the biography of Epicurus written by the ancient Greek historian of philosophy Diogenes Laertius (probably first half of the third century CE), in his *Lives of Eminent Philosophers.*

It is in the 'Letter to Pythocles' that Epicurus offers explanations of meteorological phenomena. Epicurus addresses his thoughts to Pythocles, who has asked for a summary of his views. The 'Letter' is deliberately brief; Pythocles has requested a document that he can refer to easily. It was never meant to replace Epicurus' larger works, for example the (now mainly lost) *On Nature*, but to serve as an *aide-mémoire.* In discussing meteorological phenomena and their causes, Epicurus makes it clear that his primary aim is to provide peace of mind (*ataraxia*), through abolishing fear of sometimes frightening meteorological phenomena. He states that 'knowledge of meteorological phenomena, whether taken along with other things or in isolation, has no other end in view than peace of mind and firm conviction'.[8] In seeking peace of mind, he calls for the rejection of traditional ways of explaining weather, including those which affirm divine intervention.

In Epicurus' physics, and his meteorology, the gods play no causative role. Further, he rejects the use of weather signs as an indication of the gods' action, for this violates his theology. He argues that:

> the fact that the weather is sometimes foretold from the behaviour of certain animals is a mere coincidence in time. For the animals offer no necessary reason why a storm should be produced; and no divine being sits observing when these animals go out and afterwards fulfilling the signs which they have given. For such folly as this would not possess the most ordinary being if ever so little enlightened, much less one who enjoys perfect felicity.[9]

The Epicurean gods are too busy being blissful to bother with human events, including aiding weather prediction. This direct denial of the causal link between the gods and weather signs is pointed and suggests the popularity of such signs in Epicurus' time, and the fear that some weather events evoked; even if divine wrath was not an issue, the effects on agriculture could be devastating.

It was the elimination of fear and anxiety (particularly about the intervention of the gods in the natural world) which motivated Lucretius to give an account of the philosophy and physical theory of Epicurus in his didactic Latin poem *On the Nature of Things* (*de rerum natura*),[10] one of the great works of Latin literature. As part of his project, he explains various meteorological phenomena in Book 6. Lucretius discusses celestial, meteorological, and terrestrial phenomena, which are sometimes regarded as terrifying wonders, and offers rational explanations which eliminate the search for supernatural causes. Throughout his poem, Lucretius is concerned to show that the basis of the universe is material and natural. While human beings are in some circumstances tempted to posit that supernatural or divine beings, such as the gods, are responsible for the creation and workings of the world, such explanations are not necessary.[11] Once humans understand the material workings of the natural world, there is no need to fear the supernatural intervention of gods, and *ataraxia*, freedom from anxiety, is possible.

There is a passage in Lucretius' *On the Nature of Things* (Book 6) that is very close in content to the 'theological excursus' in

Theophrastus. This suggests that the passage in Theophrastus is not wholly a medieval translator's gloss, and that Theophrastus' text very likely served as a source for Epicurus, whose ideas in any case formed the basis of Lucretius' poem.[12] Lucretius (6.357–78) answers the question about the frequency of thunder in springtime (and includes a reference to autumn as well, not found in Theophrastus) and then, like Theophrastus, addresses the question of divine retribution (6.388–422), but in a far more elaborate form:

> If Jupiter and other gods, my friend,
> Shake with appalling din the realms of heaven,
> And shoot their fire where each one wants to aim,
> Why do they not arrange that when a man
> Is guilty of some abominable crime
> He's struck, and from his breast transfixed breathes out
> Hot flames, a bitter lesson to mankind?
> Why is a man of conscience free from stain
> Engulfed in flames, all innocent, suddenly
> Seized by a fiery whirlwind from the sky?
> Why do they waste their pains shooting at deserts?
> Or are they merely practising their aim
> And strengthening their muscles? Why do they allow
> The Father's bolt to be blunted on the ground?
> Why does he allow this himself, and not keep it
> For his enemies? And why does Jupiter
> Never when the sky is cloudless everywhere
> Launch bolts upon the earth and sound his thunder?
> Or does he wait until the clouds have formed
> And then himself descend down into them
> To aim his weapon from a shorter range?
> What is his object when he strikes the sea?
> Has he some grudge against the waves and all
> The liquid mass of water and swimming plains?
> And if he wants us to beware the stroke
> Why is he loth to let us see it coming?
> But if he wants to crush us unawares
> Why does he thunder from the same direction
> And put us on our guard? Why does he first
> Summon the darkness, with its roars and growls?
> And can you possibly believe he shoots
> In many directions simultaneously?
> Or would you dare to say this never happens,

Never many strikes at the same time?
In fact this often occurs, and it must be
That just as rain-showers fall in many places
So at one time fall many thunderbolts.
   Lastly, why does he wreck the holy shrines of gods
And his own glorious habitations
With hostile thunderbolt? Why does he smash
The noble images of gods, and dishonour
His own fine statues with a violent wound?
Why does he mostly strike high ground, why do we see
The signs of fire most often on the mountain tops?[13]

Having raised these questions, Lucretius offers a series of explanations for meteorological phenomena, based on Epicurean natural
philosophy. To some extent, it is the use of multiple causation as an
explanatory strategy which signals that Epicurus and Lucretius were
operating within a Theophrastean framework.

   The use of multiple causes is a hallmark of Epicurus' approach in
his 'Letter to Pythocles'. In presenting his explanations of meteorological phenomena, Epicurus states that it is not possible to understand all matters equally well. In his view, some things may be
impossible to comprehend, while the treatment of some, including
human life as well as the principles of physics, may be more clear than
that of the *meteōra*. He explains that it is not the case, with meteorological phenomena, that only one explanation is possible: rather
these phenomena 'admit of manifold causes for their occurrence and
manifold accounts'. Epicurus deliberately argues against 'being in
love' with one dogmatic explanation which, he claims, is a superstitious trap into which 'so-called' astronomers and physicists have
fallen:[14] 'when we pick and choose among [causes], rejecting one
equally consistent with the phenomena, we clearly fall away from the
study of nature altogether and tumble into myth'.[15]

   Because he advocates a number of possible causes for meteorological phenomena, it is likely that Epicurus used Theophrastus'
work. But, there are important differences between the ways in which
multiple causes are invoked in Theophrastus' *Meteorology/Metarsiology*
and in the 'Letter to Pythocles'. In the preceding chapter,
Theophrastus was shown to have held the view that, for some phenomena, a number of different causes are at work; Theophrastus
specifies which causes are responsible for particular phenomena, often
presenting a list of causes. Epicurus is not especially concerned with
correlating specific causes with particular phenomena. Rather, he is

aiming to establish the general principle that multiple causes may result in the same phenomenon, and does not provide a detailed exposition.[16]

It is important to remember that the 'Letter to Pythocles' is a summary; any comparison between it and Lucretius' poem must recognize its deliberate brevity. In outlining his approach to explaining meteorological phenomena, Epicurus emphasizes that the causes suggested and accounts offered must not contradict experience.[17] He begins his explanation with a very brief account of celestial phenomena, particularly those associated with the sun and moon, e.g. eclipses and night and day. Addressing the regularity of the orbits of celestial bodies, Epicurus asserts that these must 'be explained in the same way as certain ordinary incidents within our own experience';[18] this analogy with everyday experience is advocated throughout the 'Letter'. Epicurus explicitly states that no divine cause can be offered for the phenomena: 'the divine nature must not on any account be adduced to explain [the celestial orbits], but must be kept free from the task and in perfect bliss',[19] and not saddled with 'burdensome tasks'.[20]

Keeping in mind the fundamental point that the *meteōra* are not the work of gods, Epicurus argues that the occurrence of weather signs can be explained in more than one way. The signs in the sky which indicate future weather may be due to the mere succession of the seasons, as is the case with the signs indicated by animals. Or, weather signs may be caused by changes and alterations in the air. Neither explanation is in conflict with the observations and it is not possible to know whether the effect is due to one or the other cause.[21]

In his 'Letter to Herodotus', Epicurus presents a brief overview of his physics; there, too, in outlining the atomic theory, he rejects divine agency, while stating that 'we must recognize . . . no plurality of causes or contingency'. But, he explains that 'when we come to subjects for special inquiry, there is nothing in the knowledge of risings and settings and solstices and eclipses and all kindred subjects that contributes to happiness . . . hence, if we discover more than one cause that may account for solstices, settings and risings, eclipses and the like . . . we must not suppose that our treatment of these matters fails of accuracy, so far as it is needful to ensure our tranquillity and happiness'.[22] Meteorology thus seems to qualify as a subject of special inquiry. Furthermore, the achievement of *ataraxia* overrides the desire for a single explanation. Epicurus justifies his advocacy of multiple causation by his explanation that 'we do not seek to wrest by force what is impossible, nor to understand all matters equally well, nor make our treatment always as clear as when we discuss human life or explain the principles of physics in general'.[23] Meteorology requires a set of methodological procedures different from those applied to general physical questions or human life.

In Epicurus' view, agreement with the phenomena is imperative; even though meteorological phenomena 'admit of manifold causes for their occurrence and manifold accounts, none of them contradictory of sensation, of their nature'. He emphasizes that 'in the study of nature we must not conform to empty assumptions and arbitrary laws, but follow the promptings of the phenomena'.[24]

But he acknowledges the difficulty of explaining some meteorological phenomena. In the case of distant ones, for example those that occur high above us, he presses the analogy with everyday experience. He argues that 'we must give explanations about the events in the heavens and everything that is nonapparent by comparing in how many ways a similar thing happens in our experience'.[25] He elaborates, explaining that 'some phenomena within our experience afford evidence by which we may interpret what goes on in the heavens';[26] so everyday experiences provide data which can be used, by analogy, to explain more distant phenomena, especially meteorological events. Epicurus, like Aristotle and Theophrastus before him, found analogy a useful tool for explaining phenomena whose causes cannot be directly observed; he explicitly advocated its use, not only in the 'Letter to Pythocles', but in his 'Letter to Herodotus' as well.[27]

In the 'Letter to Pythocles', Epicurus discusses these: clouds, rain, thunder, lightning, thunderbolts, prēstēr, earthquakes, wind, hail, snow, dew, frost, ice, rainbows, haloes, comets and falling stars. For the most part, his discussion is very brief, incorporating multiple causes and analogies to ordinary experience. So, for example, 'thunder may be due to the rolling of wind in the hollow parts of the clouds, as it is sometimes imprisoned in vessels which we use; or to the roaring of fire in them when blown by a wind, or to the rending and disruption of clouds, or to the friction and splitting up of clouds when they have become as firm as ice'.[28] Throughout, he deliberately emphasizes multiple causes, noting that the phenomena invite us to give a plurality of explanations. He sees the use of analogy as an important part of his overall method, suggesting that it is easy to see that lightning may occur in a number of ways 'so long as we hold fast to the phenomena and take a general view of what is similar to them'.[29] He goes on to explain that 'exclusion of myth is the sole condition necessary; and it will be excluded, if one properly attends to the facts and hence draws inferences to interpret what is obscure'.[30]

Notably, Epicurus' treatment of ice is markedly different from that of other phenomena, for here he makes special reference to atomic theory, and uses geometrical language (circular, scalene, acute-angled) to describe possible shapes of ice atoms:

> Ice is formed by the expulsion from the water of the circular, and the compression of the scalene and acute-angled atoms contained in it; further by the accretion of such atoms from without, which being driven together cause the water to solidify after the expulsion of a certain number of round atoms.

The use of geometrical terms to describe ice contrasts with the homely language of everyday experience (e.g. the description of wind trapped in a jar) used to describe other phenomena. But even here Epicurus provides different possibilities. (It is worth noting that Lucretius (6.527ff.) did not follow Epicurus' lead in his discussion of ice. Rather, he dismisses the other forms of matter that originate and grow and condense in clouds, namely, snow, wind, hail and frost, by noting that 'it is easy enough to discover and picture mentally how one and all come into being or are created, when once you have rightly grasped the properties of the elements'. Lucretius offers no further explanation.)[31]

While the relationship of Epicurus' views and methods to those of other predecessors is not entirely clear, Theophrastus' *Physical Opinions* does seem to have been a major influence on him and, through him, on Lucretius. Theophrastus' influence on Epicurean meteorology may be most clearly recognized in certain styles of explanation adopted by Epicurus and Lucretius, and notably the use of multiple causes and analogy. The positing of a number of possible causes may owe something to the doxographical style of collecting and presenting the opinions of others. Doxography, and the opinions of earlier natural philosophers, very likely provided the foundation for the multiple explanations offered by Epicurus in the 'Letter to Pythocles'. Sedley has argued that Epicurus used Theophrastus' *Physical Opinions* as his source for earlier views, providing examples of possible explanations.[32]

Yet there are some noticeable differences between Epicurus' approach and that of Theophrastus, who offers multiple causes as explanations only when it appears that he cannot decide on just one. This seems to be the case either because of the cause itself or because he cannot muster sufficient evidence. (In places, his calls for further research only make sense if further evidence would help to decide between possible causes.) Epicurus and his followers seem happy to entertain plural causes on any and every occasion (in aetiology at least), because they conceive of their role as providing a natural cause, so that peace of mind will follow.[33]

There is also a lack of complete correspondence between the phenomena considered by Theophrastus and Epicurus.[34] Of course, it is possible that the versions of Theophrastus' *Meteorology/ Metarsiology* to which we now have access differ from the one with which Epicurus was familiar. Nevertheless, Epicurus not only considers the phenomena in a different order from that of Theophrastus, but also discusses some not treated by him in the extant parts of the treatise, namely, eclipses, comets and falling stars. It might be argued that Epicurus' treatment of these lies outside the range of meteorological phenomena accepted by Theophrastus (or that these sections were missing from Theophrastus' work).[35] While that might be the case, comets and shooting stars were considered by Aristotle in his *Meteorology*. Futhermore, Epicurus does discuss the rainbow, which was treated at length and in some detail by Aristotle. However, one of Epicurus' explanations of the rainbow (for he characteristically offers several) is distinctly non-Aristotelian, when he invokes the aggregation of the atoms as a possible cause; neither Aristotle nor Theophrastus would have endorsed an atomistic explanation.

It is clear that Epicurus' philosophy provided the inspiration and basis for Lucretius' poem *On the Nature of Things*. The work has six books. The first two set out the physics of the universe; in the first, Lucretius describes the nature of the material constituents of the universe, atoms and the void, while the motion of the atoms in the void is explained in Book 2. The third and fourth books focus on the material composition of the human mind and spirit (Book 3) and the material basis of sensation (Book 4). The origin of the world and the growth of human society are treated in Book 5; finally, in Book 6 (which may be unfinished) Lucretius turns to meteorology.[36]

It begins with an introduction (6.1–42) in which Lucretius sings the praises of the 'fruits' of Athens: grain and her native son, Epicurus:

> He proved that mankind mostly without cause
> Stirred up sad waves of care within their breasts.
> For we, like children frightened of the dark,
> Are sometimes frightened in the light – of things

No more to be feared than fears that in the dark
Distress a child, thinking they may come true.

Lucretius outlines his purpose, to dispel mankind's irrational fears:

Therefore this terror and darkness of the mind
Not by the sun's rays, nor the bright shafts of day,
Must be dispersed, as is most necessary,
But by the face of nature and her laws.[37]

He then turns to the main subject of Book 6, providing natural and rational explanations for those phenomena which humans most commonly attribute to supernatural causes: thunder, lightning, thunderbolts, waterspouts (*prēstēr*), earthquakes, volcanoes, the flooding of the Nile and magnets. The book concludes with a description of epidemic and plague in Athens. (The order in which Lucretius discusses the meteorological phenomena does not follow the order in the 'Letter to Pythocles' at all closely; he was very likely relying on Epicurus' now-lost work *On Nature*.)[38]

In the poem, Lucretius uses an analogy to explain how multiple explanations can apply to those phenomena which are remote (6.703–11):

There is also a number of things for which
It is not enough to state one cause; we must
Consider many, and one of them is right.
For example, if from a distance you should see
The lifeless body of some man, then all
The causes of death you might think well to mention,
So that the one true cause of it be named.
For though you could not prove that steel or cold
Had caused his death, or disease perhaps, or poison,
We know quite well that what has happened to him
Is something of this kind. And so we shall
In many cases argue in this way.[39]

Lucretius shows that distant appearances which need to be explained are only sufficiently observed to allow them to be identified as instances of a general category, such as 'death'. Distant phenomena cannot necessarily be recognized as specific types within the general category. For this reason, all possible explanations must be cited if they are to include the one relevant to the particular instance. All of the explanations are potentially true, even though only one is true for each particular event.[40] The multiple causes of, for example, thunder are alternatives with regard to a particular individual thunderclap, but are all, nonetheless, possible causes of thunder? This is in contrast to the explanation of (at least some) astronomical phenomena, which have only one explanation in our world, but others in other worlds (Lucretius 5.509–33).[41]

While Epicurus had advocated the use of analogies with everyday experience in the explanation of meteorological phenomena, he had only provided sketchy examples in his summary 'Letter to Pythocles', such as the analogy between thunder being due to the rolling of wind in the hollows of clouds and wind being trapped in ordinary vessels.[42] Lucretius provides more detailed analogies. So, for example, in his explanation of thunder being caused by the wind's power splitting a cloud with a terrible crash, he notes that this is no wonder, since a small bladder filled with air makes a noise when it bursts suddenly.[43]

Lucretius offers several explanations of thunder, including a description of the noise that sometimes accompanies lightning:

> Thunder comes also when a flaming stroke
> Of lightning falls from a cloud upon a cloud.
> If the receiving cloud is full of water
> It makes a great noise quenching it at once,
> As red-hot iron taken from the furnace
> Hisses when plunged into a tank of water.[44]

He uses another detailed analogy to explain why we see lightning before we hear the thunder:

> Things always come more slowly to the ears
> Than to the eyes; as this example shows:
> If in the distance you observe a man
> Felling a tall tree with twin-bladed axe
> You see the stroke before the sound of it
> Reaches your ears; so also we see lightning
> Before we hear the thunder, which is produced
> At the same time as the fire, and by the same cause,
> Born of the same collision of the clouds.[45]

The poetic medium makes the analogies pointed to by Theophrastus and Epicurus more immediate and the explanations more comprehensible.

In *On the Nature of Things*, Lucretius elaborates Epicurean ideas. But the poem is not simply a translation from philosophical prose to didactic poetry, nor simply from Greek to Latin. Indeed, in many ways Lucretius was aware that direct translation from one culture to another is not possible. At the end of the proem to the poem (1.136–9), he apologizes for the difficulties faced in the translation:

> Nor do I fail to appreciate that it is difficult to illuminate in Latin verse the dark discoveries of the Greeks, especially because much use must be made of new words, given the poverty of our language and the newness of the subject matter.[46]

But it was due to the power and the beauty of Lucretius' poem that many readers, in other more distant cultures and times, encountered Epicurean meteorology.

The Stoics and Marcus Manilius

Another group of Hellenistic philosophers, known as the Stoics, competed with the Epicureans in the philosophical marketplace. The Stoic world-view tended, in contrast to that of the Epicureans, to emphasize the permeation of the cosmos by the divine.[47] Our knowledge of the natural philosophy of the early Stoics, including Zeno (of Citium in Cyprus, 335–263 BCE), the founder of the school, and his successors Cleanthes (of Assos in Asia Minor, 331–232 BCE) and Chrysippus (of Soli in Cyprus, *c.* 280–207 BCE), is based on reports preserved in later writers. Posidonius (of Apamea in Syria, *c.* 135–*c.* 51 BCE) wrote a work on meteorology, but, as with his many other writings, only fragments and reports survive. Diogenes Laertius, in his *Lives of Eminent Philosophers*, provides a brief and rather matter-of-fact report of Stoic meteorology in his biography of Zeno, as part of his discussion of Zeno's natural philosophy. He claims to report the views of Zeno, referring to his work *On the Whole*, and the views of Chrysippus. He also refers to two works by Posidonius, the *Meteorology* and the *Physical Discourse*. The meteorological explanations attributed by Diogenes Laertius to the Stoics are presented in a straightforward and simple manner, with no comment or elaboration, except occasionally to name a source. So, for example, winds are streams of air named according to the place from which they blow; hail is frozen cloud, having been crumpled by wind. Lightning is the kindling of clouds, as described by Zeno.[48]

The poet Marcus Manilius (early first century CE), author of the *Astronomica*, was influenced by Stoic philosophy.[49] Manilius believed that the divine spirit pervades the entire universe; the universe reflects the *ratio* (reason) of the divine will. The unity of the natural world, and its dependence on the divine spirit, is his underlying theme. While the *Astronomica* appears to reflect Stoic ideas regarding the divinity of the universe, Manilius was also clearly familiar with Lucretius' *On the Nature of Things*. In fact, it is possible that Manilius' poem was intended, in part, as an attack on Lucretius' Epicurean ideas. Manilius' insistence on the role of the divine spirit at work actively throughout the universe contrasts strongly with Lucretius' world, in which the gods have no active role.[50] But Manilius' explanation of meteorological phenomena does share something with that of Lucretius, in that these potentially terrifying phenomena are all part of the natural order; their causes can be understood by humans and therefore need not be feared.

For Manilius, the universe is a living thing, requiring nourishment provided by the divinity. The poet proclaims the unity of the cosmos (2.60ff.):

> I'll sing how God the World's Almighty Mind
> Thro' All infus'd, and to that All confin'd,
> Directs the Parts, and with an equal Hand
> Supports the whole, enjoying his Command:
> How All agree, and how the Parts have made
> Strict Leagues, subsisting by each others Aid;
> How All by Reason move, because one Soul
> Lives in the Parts, diffusing thro' the whole.
>
>   For did not all the Friendly Parts conspire
> To make one Whole, and keep the Frame intire;
> And did not Reason guide, and Sense controul
> The vast stupendous Machine of the whole,
> Earth would not keep its place, the Skies would fall,
> And universal Stiffness deaden All;
> Stars would not wheel their Round, nor Day, nor Night,
> Their Course perform, be put, and put to flight:
> Rains would not feed the Fields, and Earth deny
> Mists to the Clouds, and Vapors to the Sky;
> Seas would not fill the Springs, nor Springs return
> Their grateful Tribute from their flowing Urn:
> Nor would the All, unless contriv'd by Art,
> So justly be proportion'd in each part. . . .[51]

In their turn, the rains, winds, seas and rivers participate in the process of nourishment, feeding other parts of the universe. Manilius is concerned to emphasize the unity of nature, the links between the celestial, the terrestrial and even the subterranean. Traditionally treated as part of meteorology, rain, winds, seas and rivers play a crucial role in nourishing the living cosmos. The images of the cosmos as a living creature and of its feeding go beyond any of the descriptions in the writings of Aristotle.[52]

Manilius opens the poem with a brief history of human interest in the sky, providing an introduction to cosmogony, cosmology and astronomy. After briefly reviewing a number of theories regarding the origin of the universe and its composition, Manilius discusses the order and arrangement of the cosmos, beginning with the zodiac and constellations, moving on to the planets and the celestial circles. He then turns to a discussion of comets and meteors (1.809–926).

Manilius is not concerned with providing a single account of the causes of comets and other *meteōra*; rather, he offers several explanations to account for those fires (*ignes*) which occur on rare occasions and are then swept away (1.813–14). His willingness to include several possible causes suggests further links to Lucretius, Epicurus and Theophrastus.[53] While Manilius acknowledges that 'in times of great upheaval rare ages have seen the sudden glow of flame through the clear air and comets blaze into life and perish' (1.815–16),[54] he emphasizes that such occurrences may be explained by three possibilities, which all conform to nature and the order of the cosmos. The first explanation is that the earth, through its vapours, supplies the seeds for comets:

> Maybe the earth breathes forth an inborn vapour, and this damper breath [*spiritus*] is overpowered by an arid air; when clouds are banished for long periods of clear weather so that air grows hot and dry under the rays of the sun, fire then descends and seizes its apt sustenance [*alimenta*] and a flame takes hold of the matter that suits its nature; and since there is no solid body, but only the wandering elements of the breezes, tenuous and most like to drifting smoke, the action is short-lived and the fires last no longer than the moment of their beginning: the comets perish as they blaze.[55]

Elements of this explanation are by now familiar, having been met in other authors. Manilius' linking of comets to earthy exhalations seems to echo Aristotle (cf. *Meteorology* 1.341b5ff.), as do his images of breathing and nourishment. The words used by Manilius and Aristotle (1.341b19, 1.344a31) to refer to the flammable substance from which comets arise both carry a sense of being food and supplying sustenance (*alimenta* and ὑπέκκαυμα).

Manilius' description of the various types of comets is based on his explanation that, 'since all the drier exhalation from the earth is spread abroad and catches fire in no uniform manner', we see comets of different shapes. He then lists the different forms of comets (1.835–51), appearing as long tresses of hair or fiery beards, and as beams, columns, casks, goats, torches and shooting stars.[56]

Manilius explains that fire is present everywhere in the universe, and is linked to many natural phenomena, including some which are meteorological: '[fire] dwells in the laden clouds and forges lightning, it makes its way into the earth and threatens heaven with the flames of Etna, it causes waters to boil at their very sources, and finds a habitation in hard flint and in verdant bark, when trees dashed against trees are set aflame; to such an extent does all nature abound with fire'. Because of the ubiquity of fire, the poet urges his readers to

> wonder not that torches suddenly burst forth from the skies; and that the air is kindled and shines with flickering flames after embracing the dry seeds exhaled by the earth, seeds which the swift fire, as it feeds, both pursues and shuns, for you see the lightning hurl its quivering flash from the midst of a rainstorm and the heavens rent with the thunderbolt.[57]

But, Manilius explains, it may be the case that dim stars are attracted to and then released by the sun (1.867–73). Further, comets may serve as warnings of impending disasters:

> Death comes with those celestial torches, which threaten earth with the blaze of pyres unceasing, since heaven and nature's self are stricken and seem doomed to share men's tomb. Wars, too, the fires portend, and sudden insurrection, and arms uplifted in stealthy treachery.[58]

Manilius does not choose between the several possible explanations of comets, instead he describes the various disasters and wars presaged by the signs in the sky. But even when comets act as portents or warnings, for Manilius they are a natural part of the universe, linked to the elements of earth, air and fire.

While his discussion of meteorological phenomena, particularly comets, plays only a small role in the work, he is anxious to emphasize that comets are part and parcel of the natural world (shaped by the divine spirit). Although his emphasis on the role of the divine spirit in the natural world is in sharp contrast with Lucretius' view, there are shared elements in their meteorological explanations, notably the emphasis on the natural character of the *meteōra* and the use of multiple causation as an explanatory tactic.

Seneca

The Roman Stoic Lucius Annaeus Seneca (b. 4 BCE–1 CE, d. 65 CE) was prolific, the author of many works still extant, including tragedies, a satire and philosophical treatises. During his lifetime he was politically active and, at times, influential; for example, he was Nero's tutor. His writings cover a wide range of topics, but his primary aim was to inspire ethical improvement.[59]

He retired from public life in 62, and focused his attention on writing and philosophy. It was during this period, towards the end of his life (prior to his forced suicide in 65) that Seneca wrote the *Natural Questions*, in which he provided a detailed discussion of meteorological phenomena; the work does not survive in its entirety.[60] It is also difficult to determine the original order of the books and their topics.[61]

Several of the extant books contain prefaces and epilogues in which Seneca is concerned with moral exhortation; in this regard, the *Natural Questions* fits in with his overall programme of moral improvement. Some modern scholars have suggested that moral exhortation is really the preoccupation and primary theme of the *Natural Questions*; it has even been suggested that Seneca, in choosing such a potentially boring topic as meteorology, was demonstrating literary virtuosity.[62] But, while meteorology may not intrigue everyone, there are indications that Seneca had a long-standing interest in it; he mentions in the *Natural Questions* (Book VI) that he had as a young man written a treatise, now lost, on earthquakes. In the year 41, Seneca told his mother (*To Helvia on Consolation* = *Ad Helviam Matrem* 20.2) that he was particularly interested in the subject-matter of what later became the *Natural Questions*. While his meteorological treatise may have been part of a larger programme designed to aid in ethical improvement, Seneca's specific interest, even fascination, in meteorology seems clear.

The *Natural Questions* is addressed to his close friend Lucilius who, in turn, has sometimes been identified as a possible author of the *Aetna*, a poem concerned with volcanic activity.[63] Seneca and Lucilius very likely shared a common interest in meteorological (especially seismological) phenomena, which they pursued in later life, as well as when they were younger.

The *Natural Questions* is divided into separate books that centre on, for the most part, meteorological topics.[64] While there is certainly a strong moral content and focus to the work, the detail in which the phenomena are explored argues against the view that the meteorological theme is merely a vehicle for Seneca's ethical messages. In contrast, Epicurus and Lucretius both appear to have dealt with meteorology primarily as a means to the ethical end of achieving *ataraxia*; their writings on the topic do not suggest a particular interest in meteorology. Their focus on meteorological phenomena is deliberate, insofar as those phenomena are especially frightening to many people. But, by eliminating fear of them, Epicureans could go a certain way towards *ataraxia*. Seneca too, as is clear in various parts in the *Natural Questions*, was concerned with alleviating fear. Yet, on his own admission, he seems to have been genuinely interested in meteorology for its own sake, and particularly earthquakes.

His approach to explaining meteorological phenomena shares much with his predecessors. He discusses and criticizes, at great length, others' explanations, in the course of presenting his own views and arguing for his own ideas. Like Aristotle's *Meteorology*, Seneca's *Natural Questions* is useful in that it provides information on the history of the subject. As part of his approach, he presents a great deal of material on earlier thinkers, often presenting an overview of their ideas and theories. He mentions many thinkers whose works are lost to us. In his discussion of comets, he focuses on the theories of Epigenes and Apollonius of Mynda, authors of whom we otherwise know very little. Seneca is also the source of much of what we know about the meteorological ideas of the Stoic philosopher Posidonius (*c.* 135–*c.* 50 BCE).[65]

In the *Natural Questions*, Seneca addresses the main topics of ancient meteorology, and covers (with very few exceptions) the same subjects as those covered by Aristotle in Books 1–3 of his *Meteorology*, although Seneca does not discuss the Milky Way or the sea.[66] Seneca also uses many of the same explanatory tactics as his predecessors. Like Aristotle, Theophrastus, Epicurus and Lucretius, he employs analogies with everyday experience; some of those he introduces are particularly homely, and offer details about contemporary Roman life. He was particularly interested in providing clear definitions of terminology, and in discussing etymology. He touches only briefly on the use of 'experiment', and mentions mathematical approaches almost disparagingly. He also provides a rather detailed, and somewhat historical, discussion of signs and auspices.

He opens the *Natural Questions*[67] with a greeting to his friend Lucilius and an explanation of his overall project, which is 'to survey the universe, to uncover its causes and secrets, and to pass them on to the knowledge of others'. He derides the efforts of historians, concerned with what has already happened, arguing that 'it is far better to investigate what ought to be done rather than what has been done . . .'.[68] He notes that it is helpful to study nature, for a number of reasons: to abandon sordid matters, to free the mind from the body, and to exercise the mind on the mysteries of nature, in order to tackle ordinary problems successfully.[69]

Seneca begins with the subject of rivers.[70] Here, he presents some of the basic principles which underlie his understanding of nature, such as the idea that 'all elements come from all others: air from water, water from air, fire from air, air from fire'. He explains that:

> there are reciprocal exchanges on the part of all the elements. What is lost to one passes over into another. And nature balances its parts just as though they were weights on scales, lest the equilibrium of its proportions be disturbed and the universe lose its equipoise.[71]

His use of the four elements is fairly standard, and the idea of 'balance' is reminiscent of Theophrastus.[72]

In his discussion of rivers, Seneca draws an extended analogy between the earth and the human body; this is invoked in other books as well. He explains that 'in the earth also there are some routes through which water runs, some through which air passes. And nature fashioned these routes so like human bodies that our ancestors even called them "veins" of water [*aquarum venas*].' Seneca goes on to describe the kinds of moisture present in human bodies (blood, brains, bone marrow, mucus, saliva, tears and the lubricant of the joints), and those found in the earth, noting that 'just as in our bodies, the liquids in the earth often develop flaws':

> Therefore, just as in our bodies when a vein is cut the blood continues to ooze until it all flows out or until the cut in the vein has closed and shut off the bleeding, or some other

cause keeps the blood back; so in the earth when veins are loosened or broken open a stream or a river gushes out.[73]

His drawing out of the analogy between the earth and the human body is elaborate and vivid, and more detailed than that found in Aristotle's *Meteorology* or Manilius' *Astronomica*.

At certain points Seneca's descriptions of natural phenomena are strikingly visual and dramatic; the final portion of the book is concerned with the impending deluge (analogous to the periodic fiery conflagration posited by the Stoics; cf. III.28.7). Seneca's description of the various stages leading up to the deluge are extremely vivid. But the potential effect of water on humans is not simply calamitous; near the beginning of the book (*NQ* III.1.2) he mentions the medicinal usefulness of certain river waters.

Other analogies and images are used to describe the nature of rivers. In recounting Empedocles' theory concerning hot springs, Seneca draws an analogy with contemporary Roman plumbing, noting that:

> We commonly construct serpent-shaped containers, cylinders, and vessels of several other designs, in which we arrange thin copper pipes in descending spirals so that the water passes round the same fire over and over again, flowing through sufficient space to become hot. So the water enters cold, comes out hot. Empedocles conjectures that the same thing happens under the earth.[74]

The Nile is dealt with in a separate book, for it is unique. Egyptian farmers operate quite differently from Romans, because they depend on the flooding of the Nile; they have no need to look towards the sky.[75] Seneca names a number of authors, dramatists and natural philosophers, as sources of information about the Nile.[76] He notes that Aeschylus, Sophocles and Euripides each reported that the rising of the Nile in summer is due to melted snow from the mountains in Ethiopia.[77]

In what remains of the following book (Book IVB), Seneca discusses hail and snow; sections on clouds and rain may have come first but are now missing. He suggests a simple distinction between hail and snow as being, in the first case, suspended frost and, in the second, suspended ice. Here, as elsewhere in the *Natural Questions*, Seneca strikes a personal note, echoing his bond of friendship with Lucilius; often, as here, he is deliberately jocular:

I could stop now, since the investigation is complete, but I will give you your money's worth; and since I have already started to be a nuisance to you I will discuss everything asked about this subject.[78]

Seneca describes a number of theories about hail, continuing in this somewhat joking fashion, suggesting that it is sensible to present even theories with which he finds fault, for most have some problem:

I am afraid either to mention or to omit a theory established by our Stoic friends, because it seems without basis. But what is wrong in writing about it to so indulgent a judge as you? Actually, if we started to subject every argument to a rigid test we would be compelled by command to silence. There are only a few theories without opposition; the others are pleading in court, even though they eventually win their case.[79]

With this warning, Seneca explains that 'there are some experts in observing clouds and in predicting when hail will come. They are able to know this just from experience, since they have noted in the clouds a colour which is in every case followed by hail.'[80] He continues, expecting Lucilius to smile at this account:

It is incredible that at Cleonae there were 'hail-officers' appointed at public expense who watched for hail to come. When they gave a signal that hail was approaching, what do you think happened? Did people run for woolly overcoats or leather raincoats? No. Everybody offered sacrifices according to his means, a lamb or a chicken. When those clouds had tasted some blood they immediately moved off in another direction.[81]

He asks whether Lucilius is laughing at this, and adds further mocking details:

If someone did not have a lamb or a chicken he laid hands on himself, which could be done without great expense. But do not think the clouds were greedy or cruel. He merely pricked his finger with a well-sharpened stylus and made a favourable offering with this blood, and the hail turned away from his little field no less than it did from the property of a man who had appeased it with sacrifices of larger victims.[82]

Seneca explains that in ancient times people used sacrifices and incantations to try to control the weather. Even the Twelve Tables,
traditionally regarded as the earliest written form of Roman legal
statutes, contained the warning: 'No one may make incantations
against another's crop.' But, now, 'it is so obvious none of these things
can happen that it is unnecessary to enrol in the school of some philosopher to learn it'.[83] This may be a veiled criticism of the Epicureans.

Occasionally, Seneca suggests conducting a sort of 'experiment' to
test a theory or confirm an analogy. He describes the view that cold
air rises from frozen ground in the spring and mixes with warmer air.
He recommends that those who hold this opinion 'add a proof which
I have never tested and do not plan to test (but you, I think, might
test snow at your own risk if you want to find out the truth); they
say that feet feel less cold when they step on hard, solid snow than on
soft, slushy snow'.[84] This is just part of his joking with Lucilius, but
elsewhere in the *Natural Questions* he stresses the importance of experience as a test of ideas. He uses his own experience, for example, when
he states that 'as a diligent vine-gardener myself I assure you that no
rainfall is so heavy it wets the ground to a depth beyond ten feet'.[85]

Winds are the subject of Book V, which begins not with a pref-
ace, but with a brief discussion of the proper definition of wind: wind
is flowing air (*aer*). Seneca amplifies upon this immediately, explaining that he finds more accurate the definition of wind as air flowing
in one direction (V.1.1). His concern with definition is reminiscent
of Aristotle's and Theophrastus' approach to explanation; they also
strove to define their terms.[86] Seneca then turns to consider various
theories of the winds. In his discussion of Democritus' theory, he
likens the motion and interaction of people in the marketplace to the
motion and interaction of atoms:

> as long as only a few people are in a market-place or on a
> side-street they walk about without disturbance, but when
> a crowd comes running together into a narrow place people
> bump into each other and start quarrelling. The same thing
> happens in the space surrounding us. When many bodies
> [*multa corpora*] fill a tiny space they unavoidably knock
> against each other and push, are shoved back, entwined, and
> squeezed. Winds are produced from these activities.[87]

It is not clear whether this analogy is derived from Democritus' own
account, or is Seneca's illustration of Democritus' ideas. In any case,
Seneca rejects Democritus' explanation of the production of winds;

his own view is that winds are formed in more than one way. As one possibility, he suggests (at V.4.1) that 'sometimes the earth itself ejects a great quantity of air, which she breathes out from hidden recesses'.

His discussion of this cause (the expulsion of air from hidden recesses within the earth) leads him to consider a theory which 'I cannot persuade myself either to believe or to remain silent about'. He explains that:

> In our bodies flatulence is caused by food. This flatulence is not emitted without great offence to our sense of smell – sometimes it unburdens the stomach noisily, sometimes more discreetly. In the same way, people think, this great nature of the world emits air in digesting its nourishment. It is lucky for us that nature always digests thoroughly what she consumes, otherwise we might fear a more offensive atmosphere.[88]

This analogy between the 'digestive' processes of the earth and the human body should by now be familiar, for in several parts of the *Natural Questions* Seneca employs analogies between the earth and the human body. Aristotle (2.8.366b15–20) had also explained earthquakes as being similar to flatulence.[89]

After a general discussion of the nature of wind, Seneca, beginning in chapter 7, considers winds individually, specifying where they come from and when. As part of this discussion, he presents various systems of classifying the winds and comments on the use of technical terms (often from Greek).[90] So, for example, he draws an analogy (V.17.1) between the classification of winds and the inflexion of nouns into cases. Just as not every noun has examples of all six cases,[91] so there are not always examples of all twelve winds to be found in every location.

Seneca makes the point that some winds are found only in certain localities. He gives specific examples and remarks that 'I would have an infinite chore if I wanted to discuss each and every wind. Practically every place has some wind that rises from it and falls only there.'[92] This sense of meteorological conditions particular to a specific locality is reminiscent of the Peripatetic works on winds.[93] Seneca ends his discussion with a detailed consideration of the usefulness of winds for navigation and commerce, and the potential harm which can be brought through the destructive use of such knowledge, for example in sailing to foreign lands to wage war.[94]

147

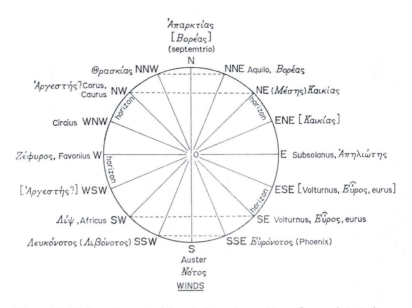

Figure 4.1 Modern diagram of the winds at the earth's surface, and their directional sources.

Source: Seneca *Naturales Quaestiones*, trans. T.H. Corcoran (1972) Cambridge, Mass.: Harvard University Press, *Loeb Classical Library*, facing 2: 312.

Note: See also E.H. Warmington's 'Note on Winds' in Seneca *Naturales Quaestiones*, (*LCL*) 2: 311–12.

The more general meteorological significance of wind is taken as part of his consideration of earthquakes and their causes. For Seneca, moving air, wind, is an important cause of seismic activity. He begins Book VI by explaining to Lucilius that he has just heard about the earthquake that struck Pompeii in 62/63. He focuses on the fear that such natural disasters invoke:

> We marvel at none of these phenomena without fear. And since the cause of fear is ignorance, is it not worth a great deal to have knowledge in order not to fear? It is much better to investigate the causes and, in fact, to be intent on this study with the entire mind. For nothing can be found worthier than a subject to which the mind not only lends itself but spends itself.[95]

He proposes to investigate the causes of such phenomena. Yet we sense that Seneca's motive is not only a desire to alleviate fear, for he

Figure 4.2 Marble wind-rose (anemoscope), *c.* 200 CE, found south of Porta
Capena, near the Via Appia in Rome. Diameter: 55 cm. Twelve-
wind diagram engraved on cylindrical marble block, with central
hole for a pole supporting a pennant and small holes near the rim
for wooden pegs indicating the winds. Dilke (1987) suggests that
'the anemoscope must have been intended as a meteorological device,
partly to help the traveler who, as he set out from Rome on the
Via Appia, would be facing south as the map does. The flag would
show the name, origin, and direction of the wind.' Cf. this marble
wind-rose, utilizing pegs, to the stone *parapēgmata* discussed in
chapter 2.

Source: Biblioteca e Musei Oliveriana, Pesaro.
Note: See also Figure 5.4, for another view.

Figure 4.3 Marble relief depicting the earthquake of 62/3 CE in the area of the Civic Forum at Pompeii, dated to 62–79 CE and found in the atrium of the House of L. Caecilius Iucundus.

Source: Soprintendenza Archeologica di Pompei.
Note: See Ciarallo and De Carolis (1999) 68.

explains that 'the investigation of [these phenomena] is so appealing to me that, even though at one time as a young man I published a volume on earthquakes, none the less I would wish to test myself and find out whether age has added anything to me in the way of knowledge or at any rate in the way of diligence'.[96] It gives him pleasure to work through the familiar material again.

He first considers the ideas of Thales, whose emphasis on water as a cause he rejects, as he does the ideas of others pointing to water as the underlying cause. He then presents the view of Anaxagoras, that fire is the cause, rejecting this theory as well and its elaboration by others.[97] He explains that 'it is a favourite theory of most of the greatest authorities that it is moving air which causes earthquakes',[98] and names Archelaus (fl. fifth century BCE), Aristotle, Theophrastus and Strato of Lampsacus (died c. 269 BCE) as holding some form of this theory.

As he relates the views of others, Seneca provides examples of the use of analogy and the acceptance of multiple causation in meteorological explanations. So, for instance, Seneca explains that Democritus offered multiple explanations for earthquakes, as sometimes caused by wind, sometimes by water, sometimes by both.[99] He relates the ideas of Metrodorus of Chios (fourth century BCE) about earthquakes, though he disagrees with him. According to Seneca, Metrodorus explained that

> when someone sings into a large jar his voice vibrates and runs through the whole jar and resonates with a kind of quavering. Even though the voice is projected only slightly it none the less travels around and causes a jolting and disturbance in the surrounding jar. In the same way the vast caves hanging down under the earth have air of their own, which other air, as soon as it falls from above, strikes and agitates, the way those empty spaces . . . just mentioned vibrate when a shout is sent into them.[100]

Here, Seneca provides another example of the common use of everyday analogies to explain meteorological events.

Seneca notes that some who believe that earthquakes are caused by air do not accept Aristotle's theory, but press an analogy between the earth and the human body. He outlines this view, in which there are passageways for water and air in the earth, similar to the veins in our bodies. He elaborates: 'in our body the movement of the veins also preserves its rhythm undisturbed while there is good health but when there is something wrong the movement pulses more rapidly and inhaling and exhaling give signs of effort and exhaustion'. Similarly, when something is wrong with the earth, 'then there is motion just like that of a sick body, because the air which was flowing through it in an even pattern is struck violently and causes its veins to shake'.[101]

The analogy between the earth and the body is used again when Seneca goes on to describe the nourishing power of the earth; everything in the universe draws nourishment from the earth and its exhalations.[102] (Seneca also refers to the nourishing function of the earth, in providing sustenance, at II.5.1–2.) In Book VI, he describes the great quantity of air, necessary for the sustenance of the universe, contained within the earth, in underground spaces. The analogy with the body occurs again later when Seneca argues against the notion that air within the earth can escape through its surface. This is incredible, because 'even in our bodies the skin keeps out the air, and air has no way in except by way of the parts through which it is breathed in, and even when it has been inhaled by us it cannot settle except in the relatively open part of the body'. Similarly, 'the same may be supposed about the earth from the fact also that an earthquake is not in the earth's surface or around the surface but underneath, from the depths'.[103] Obstructions to the movement of underground air are the cause of earthquakes;[104] analogous situations occur in the human body:

> which makes it obvious that an earthquake is brought about by moving air: our bodies also do not tremble except when some cause disturbs the air inside, when it is contracted by fear, grows weak in old age, becomes feeble with sluggish veins, is paralysed by cold, or is thrown from its normal course under an attack of disease. For, as long as the air flows without damage and proceeds in its usual way, there is no tremor in the body; when something happens which inhibits its function, then it no longer is strong enough to support what it had maintained in its vigour. As it fails it causes to collapse whatever it had sustained when it was intact.[105]

Seneca concludes the book on earthquakes with another reference to the Campanian earthquake. He explains that 'it is more important for us to be brave than to be learned. But the one does not occur without the other, for strength comes to the mind only from the liberal arts and the study of nature.'[106]

Book VII deals with comets, and opens with a brief history of astronomy. In the course of the discussion, Seneca provides many details of his, and others', astronomical views.[107] He explains that the Greeks distinguished three types of comets: 'those from which the flame hangs down like a beard, those which scatter a sort of hair around them on all sides, and those which have indeed a kind of dispersed fire but stretching to a point'.[108] After reviewing and criticizing the views of several predecessors, Seneca explains that he regards comets not as a sudden fire, but as one of the eternal works of nature (*aeterna opera naturae*). Comets 'move, preserve their continuity, and are uniform'. Each comet 'has its own position and so is not quickly expelled but measures out its own space. It is not extinguished but simply departs.'[109] In Seneca's view, comets are very special:

> Nature does not often display comets. She has assigned to them a different place, different periods, movements unlike the other celestial bodies. Also, she wished by means of comets to honour the magnitude of her work. The appearance of comets is too beautiful for you to consider an accident, whether you examine their size or their brightness, which is greater and more brilliant than the other celestial bodies. In fact, their appearance has a kind of exceptional distinction. They are not bound and confined to a narrow spot but are let loose and freely cover the region of many celestial bodies.[110]

Furthermore, comets can provide signs of the weather, 'in the way that the equinox is a sign of the year turning to hot or to cold'. He points specifically to the comet which appeared in the consulship of Paterculus and Vopiscus (in the year 60) as having provided a sign of the violent storms which occurred everywhere and the earthquakes in Achaia and Macedonia.[111]

Seneca then discusses (in Book I) the lights in the sky which are 'those fires which the atmosphere [*aer*] drives across the sky'. According to Seneca, 'they move obliquely at very high speeds, which is proof that they have been driven by a great force. It is obvious that they do not move on their own accord but are hurled.'[112] He discusses haloes, using analogies with everyday experience (including burst pipes at I.3.2), then turns to the problems of investigating the rainbow, beginning with a review of various theories.

Seneca takes the sun and cloud to be the basic causes of the rainbow, which he considers to be an image that is reflected as in a mirror. He explains (I.3.11) that the rainbow only ever occurs opposite the sun. While he asserts that 'no one can doubt that a rainbow is a reflection of the sun formed in a moist, hollow cloud', he warns that rainbow 'research is vague, where we have nothing we can get our hands on and mere conjectures must be applied extensively'. He presents his own view (which he claims agrees with that of Posidonius) that 'rainbows arise in a cloud shaped like a hollow, round mirror, whose form is that of a ball cut through the middle'. He cites the usefulness of mathematics in understanding the reflected nature of the rainbow, in that 'this cannot be proved without the aid of geometry, which instructs us with proofs that leave no doubt that a rainbow is a representation of the sun, but not a copy. For not all mirrors reflect faithfully without distortion.'[114] His reference to mathematical explanations is abrupt and, perhaps, ironic. He asserts that 'the proofs offered by mathematicians are not only persuasive but convincing, and no one is left with any doubt that a rainbow is an image of the sun, imperfectly reflected because of the flawed shape of the mirror'. But this very brief mention of the mathematics of the rainbow contrasts to the far more detailed treatment provided by Aristotle. Instead, Seneca resorts to other proofs, the sort that can be picked up on the street, so to speak. (Elsewhere, in a letter to Lucilius, Seneca makes rather disparaging remarks about the limited usefulness of mathematics.)[115]

Seneca's discussion of theories of the rainbow is characterized by his extensive consideration of various analogies: to dyed cloth, to feathers, but especially to mirrors. He elaborates his own experience with 'fun-house' sorts of mirrors, noting that:

> There are some mirrors you are afraid to look into: they reflect such a deformity from the distorted image of the viewer; the likeness is preserved – but made to look worse than it is. There are other mirrors which can make you pleased with your strength when you look into them. Your arms grow so much larger and the appearance of your whole physique is increased to more than human proportions. Some mirrors show the right side of the face, others the left. Others twist and invert the face.

Seneca concludes his account by asking the question posed by his analogy with everyday experience: 'why is it surprising that a mirror of this sort also occurs in a cloud and reflects a defective image of the sun?'[116] In this way, Seneca asserts that everyday experience can be useful in understanding natural phenomena. He then considers other lights in the sky, including 'streaks', 'suns' and 'fires'; some of these are considered to be signs of weather, particularly storms.[117]

Book II, the final book, on lightning and thunder, is the longest in the surviving *Natural Questions*. At its beginning, Seneca offers another glimpse of his interest in earthquakes (the subject he had treated extensively in a separate book):

> 'Why,' you ask, 'have you put the study of earthquakes in the section where you will talk about thunder and light-ning?' Because, although an earthquake is caused by a blast, a blast is none the less air in motion. Even if the air goes down into the earth it is not to be studied there. Let it be considered in the region where nature has placed it.[118]

He then presents a detailed account of the nature of the *aer*. He explains the need to provide such an account, because thunder and lightning occur in the *aer*.[119] In his view, some understanding of the *aer* is needed to explain these phenomena.

He explains that *aer* is an essential part of the universe, for it is:

> what connects heaven and earth and separates the lowest from the highest in such a way that it none the less joins them. It separates because it intervenes midway; it joins because through it there is a communication between the two.[120]

The *aer* 'transmits to the upper region whatever it receives from the earth', but also 'transfuses to earthly objects the influences of the stars'.[121] While the *aer* is part of the whole universe, 'it receives what-ever the earth sends for the nourishment of the heavenly bodies'.[122] Significantly, Seneca states that 'all the atmosphere's instability and disturbance is derived from this earthly element'.[123] But it is import-ant to understand that *aer*:

> is not the same throughout its entire expanse. It is altered by its surroundings. Its highest region is extremely dry and hot, and for this reason also very thin because of the nearness of the eternal fires, the many movements of the stars, and the continuous revolution of heaven. The lowest region, near the

earth, is dense and dark because it receives the terrestrial exhalations. The middle region is more temperate (compared to the highest and the lowest regions, as far as dryness and thinness goes), but it is colder than both the other regions. The upper regions of air feel the heat of the nearby stars. The lower regions are also warm; first because of the exhalation of the earth, which carries with it a great deal of warmth; second, because the rays of the sun are reflected back and make the air more genially warm with reflected heat as far as they are able to reach. Besides, the lower air is warmed by the breath which comes from all the animals, trees, and plants; for nothing is alive without heat.[124]

Furthermore, 'since the atmosphere is divided in such a way, it is especially variable, unstable and changeable in its lowest region. The air near the earth is the most blustering and yet the most exposed to influences since it is both agitating and being agitated. Yet it is not all affected in the same way. It is restless and disturbed in different parts and in different places.'[125] These variations in the *aer* in different places are crucial for Seneca's understanding of local meteorological variations.

The changes in the atmosphere are due to various motions in the universe. In some cases, 'the earth supplies some causes of the atmosphere's change and instability. The earth's position ... greatly influences the condition of the atmosphere.' But there are also celestial causes for changes in the atmosphere:

> The movements of the stars, of which you may consider the sun the most influential, supply other causes. The year follows the sun, and winters and summers alternate in accordance with the sun's circling changes. Next in importance is the influence of the moon. But other heavenly bodies also affect the earth and the atmosphere covering the earth. Their course or their concourse cause, in opposite ways, sometimes cold, sometimes rain, and other inclement disturbances on earth.[126]

Clearly, in Seneca's account, there are celestial causes for changes in the weather. Later, as part of the discussion of theories explaining lightning, Seneca argues that 'both the lowest *aether* has something similar to *aer* and the highest *aer* is not dissimilar to the lowest *aether*, because a transition from different to different does not occur immediately. On the confines they mingle their properties so gradually that you cannot tell whether it is *aer* or already *aether*.'[127] For Seneca, a material similarity between celestial and terrestrial regions helps to explain the influence of the celestial on meteorological phenomena.

He provides definitions to distinguish between lightning flashes, lightning bolts and thunder: 'thunder is made simultaneously with the others but is heard later. A lightning flash displays fire; a lightning bolt emits it. A lightning flash is a threat, so to speak, a feint without actually striking; a lightning bolt is an attack with a hit.'[128] But he prefaces his discussion by noting the disagreement among those who have considered these phenomena:

> There are certain points which are agreed upon by all authorities, others on which opinions vary. It is agreed that all these phenomena occur in the clouds and come from the clouds. It is further agreed that both lightning flashes and lightning bolts are either fiery or have the appearance of fire.[129]

Pointing to the lack of agreement, he signals that there is room for further investigation.

Like Theophrastus before him, Seneca reports some of the marvels associated with lightning strikes, such as the fusing of coins[130] and melting of swords; he also considers how it is that lightning serves as a sign of future events.[131] At some length, he explains that the difference between 'us' and the Etruscans is that 'we think that because clouds collide lightning is emitted', while the Etruscans believe that 'clouds collide in order that lightning may be emitted' which is as a sign for the future.[132] But the Stoics do not regard lightning purely as the product of clouds colliding; it may serve as a sign or portent because it is connected within the nexus of causes which regularly occur in conjunction. Lightning and other signs do not cause what is signified, but occur together with those phenomena which signs are taken to predict. It is because of this view of the regularly occurring conjunction of causes that Stoics can regard divination as an important example of an *empirically* based science.[133]

As part of his consideration of divination and the use of different types of auspices, Seneca explains that 'an auspice is the observer's auspice'; that is, observers choose what sort of sign they wish to notice. He then returns to his discussion of lightning, and enumerates the three relevant areas of study: 'how we investigate it, how we interpret it, how we charm it away'. He explains that 'the first area pertains to classification, the second to divination, the third to propitiating the gods; it is fitting to ask when lightning is good, to pray against it when it is bad; to ask that the gods fulfil their promises, to pray that they set aside their threats'.[134]

A lengthy discussion of fate follows, together with the assertion that a soothsayer (*haruspex*) may prove beneficial, in that he is a minister of fate.[135] Seneca then describes the classification of lightning adopted by Aulus Caecina, the first-century BCE Roman authority on divination and lightning who wrote a work on Etruscan lore known as *Etrusca Disciplina* (also used by Pliny the Elder).[136] The by-now-familiar question of a deity's role in causing thunderbolts occupies Seneca, and he devotes serious attention to the answer, which is interwoven into his detailed account of lightning.

As might be expected, Seneca (or an imaginary interlocutor) asks questions like those raised by other authors, concerning, for instance, the reasons why evil people are left untouched while innocents are struck down. Accepting that such a big issue cannot be properly addressed in the present work, he offers his own view briefly, that 'lightning bolts are not sent by Jupiter but all things are so arranged that even those things which are not done by him none the less do not happen without a plan, and the plan is his'.[137]

He argues that even the Etruscans did not think that the gods personally wielded weapons of destruction, stating that:

> The ancient sages did not even believe that Jupiter, the sort we worship in the Capitol and in other temples, sent lightning by his own hand. They recognized the same Jupiter we do, the controller and guardian of the universe, the mind and spirit of the world, the lord and artificer of this creation.[138]

Seneca explains that the Etruscans, in attributing the hurling of lightning bolts to Jupiter, had found a way to frighten people into behaving properly: 'being very wise they decided that fear was necessary for coercing the minds of the ignorant, so that we might fear something above ourselves. . . . And so to terrify men who find nothing attractive in good behaviour unless it is backed up by fear, they placed an avenger overhead, and an armed avenger at that.' So the Etruscan sages had adopted this line as a 'scare tactic'; they did not themselves believe that Jupiter personally hurls bolts of lightning.[139]

The divine nature of Jupiter may be understood in various ways, and given different names, according to Seneca:

Any name for him is suitable. You wish to call him Fate? You will not be wrong. It is he on whom all things depend, the cause of causes. You wish to call him Providence? You will still be right. It is by his planning that provision is made for this universe so that it may proceed without stumbling and fulfil its appropriate functions. You wish to call him Nature? You will not be mistaken. It is he from whom all things are naturally born, and we have life from his breath. You wish to call him the Universe? You will not be wrong. He himself is all that you see, infused throughout all his parts, sustaining both himself and his own.

This view was shared by the Etruscans, who 'said lightning was sent by Jupiter because nothing is done without him'.[140]

Seneca's very lengthy classification and interpretation of lightning reviews the ideas of others (including philosophers and the Etruscans) and presents restatements and refinements of his own (e.g. at II.57.1–4), as well as those of the otherwise unknown Attalus.[141] But he concludes his treatment of lightning, stating that 'I know what you have wanted for a long time, and what you keenly ask. You say, "I should rather I did not fear lightning than know about it. So, teach others how lightning bolts occur. I want to shake off the fear of them, not have their nature explained to me".' Seneca, like the Etruscans, is concerned with encouraging moral behaviour; but, unlike them, Seneca is aiming to dispel, not inspire, fear. He accepts the plea of his listener/reader, and asserts that:

> Some moral ought to be mixed in all things and all con-
> versation. When we go into the secrets of nature, when we
> treat the divine, the soul ought to be delivered from its ills
> and occasionally strengthened, which is necessary even for
> learned men, especially those who deal with this study exclu-
> sively, not in order that we may escape the blows of things –
> for weapons are hurled at us from all sides – but in order
> that we may endure them bravely and firmly.[142]

Seneca is unwilling to end his story, and his meteorological treatise, without an appropriate moral.

Some scholars have emphasized the ethical tone and purpose of the work. Brad Inwood has summarized the contribution of the *Natural Questions*, which offers the reader striking consolation for the fear of death; a sober analysis of the relationship between the cosmic order and human life; challenging epistemological reflections, focussing on the ambivalent nature of human knowledge in a cosmos which is rational but not fully open to our enquiring minds; and a sustained meditation on the relationship of man to a rational god, providential but disinclined to reveal the truth except through his orderly and causally determinate works.

He concluded that 'the purpose of the work is markedly different from that of other meteorological enquiries'.[143] Yet, as we have seen, other works dealing extensively with meteorology, notably Epicurus' 'Letter to Pythocles' and Lucretius' *On the Nature of Things*, also confront theological and ethical issues.[144] Like Epicurus and Lucretius, Seneca is concerned to free people from the fear associated with powerful, often terrifying, meteorological phenomena. This common aspiration of delivering people from fear, even fear of death, was a goal shared by several different Hellenistic philosophical schools; for Epicurus, Lucretius and Seneca it was an important aim of natural philosophy.

At several points in the *Natural Questions* Seneca complains that 'there is no interest in philosophy', and that 'so little is found out from those subjects which the ancients left partially investigated that many things which were discovered are being forgotten'.[145] Here too he addresses Lucilius directly, and reiterates the value of studying philosophy. He outlines the differences between the two branches of philosophy, one dealing with man, the other with the gods (I.Pref.1), and the differences between man's nature and that of god: 'In ourselves, the better part is the mind, in god there is no part other than the mind.' Humans' greatest possession is reason: 'none the less, meanwhile, a great error possesses mortals: men believe that this universe, than which nothing is more beautiful or better ordered or more consistent in plan, is an accident, revolving by chance, and thus tossed about in lightning bolts, clouds, storms, and all the other things by which the earth and its vicinity are kept in turmoil'.[146] In spite of their capacity to reason things out, humans neglect the study of the cosmos, and so fear the *meteōra*, not understanding their place in the ordered universe. Seneca reasserts the value of studying the nature of the universe, for 'having measured god I will know that all else is petty'.[147] His concern with divinity and the order of the cosmos was shared by many ancient authors, working in various traditions.

It is Seneca's detailed treatment of meteorological phenomena in the *Natural Questions* which is unusual. As Inwood has noted, 'in his old age, Seneca devoted a quite surprising amount of energy to meteorological enquiry'. He suggests that Seneca chose the particular format for his work as a literary challenge, to strengthen his claim as a writer, and not 'just' a philosopher, intimating that if Seneca could write compellingly on a topic as potentially tedious as meteorology, he would consolidate his literary reputation.[148] But given the detail with which Seneca discusses the various meteorological phenomena and their causes, and his confessed long-standing fascination with the subject, it seems likely that Seneca had a genuine curiosity about meteorology and was deeply interested in understanding the phenomena. That this long-standing interest in meteorology was specifically linked to a desire to alleviate anxiety and to further philosophical engagement is underlined in the closing words of his letter addressed to his mother, aiming to console her while he was in exile (many years before the writing of the *Natural Questions*):

> I am as happy and cheerful as when circumstances were best. Indeed, they are now best, since my mind, free from all other engrossment, has leisure for its own tasks, and now finds joy in lighter studies, now, being eager for the truth, mounts to the consideration of its own nature and the nature of the universe. It seeks knowledge, first, of the lands and where they lie, then of the laws that govern the encompassing sea with its alternations of ebb and flow. Then it takes ken of all the expanse, charged with terrors, that lies between heaven and earth – this nearer space, disturbed by thunder, lightning, blasts of winds, and the downfall of rain and snow and hail. Finally, having traversed the lower spaces, it burst through to the heights above, and there enjoys the noblest spectacle of things divine, and, mindful of its own immortality, it proceeds to all that has been and will ever be throughout the ages of all time.[149]

On the Cosmos

The work known as *On the Cosmos* (*De Mundo*), by an unknown author but attributed to Aristotle, treats meteorological topics as part of a larger discussion of the cosmos. Addressed to one 'Alexander', *On the Cosmos* is presented as a letter, a form commonly used to provide a summary of a subject. The evidence for dating is confused and difficult to interpret. Various dates have been suggested; some time between *c.* 50 BCE and 180–90 CE seems likely. While the identity of the author is not at all clear, the work itself provides persuasive evidence that it was not Aristotle. The invocation of Plato, and of the *Laws*, at the very end, seems fairly convincingly un-Aristotelian. Furthermore, its style is unlike surviving works of Aristotle. The treatise may even have been attributed to Aristotle by the author himself; the 'Alexander' to whom the work is addressed (referred to as 'the best of princes') could be read as Alexander III of Macedon ('the Great'), Aristotle's student. The author's desire to associate his work with Aristotle may have inspired the choice of addressee, so the date of Alexander's death in 323 BCE would not rule out a later date.[150]

On the Cosmos is primarily theological in flavour, urging the reader to adopt the author's view that a transcendent god maintains the order of the cosmos. The open-letter style adopted in it was used by other authors writing on meteorology. It is possible that the anonymous author of *On the Cosmos* modelled his style on the letters of Epicurus,[151] although the theology of *On the Cosmos* is radically different. Epicurean gods dwell in a place between worlds, leading a blissful existence, in no way involved in the workings of this world. In Epicurus' theology, 'the divine nature must not on any account be adduced to explain [the celestial orbits], but must be kept free from the task and in perfect bliss'.[152] Just as Manilius' poem the *Astronomica* may have been, in part, a Stoic-inspired rebuttal of Lucretius' Epicurean theology in his poem *On the Nature of Things*, so *On the Cosmos* looks like a Peripatetic refutation of Epicurus' 'Letter to Pythocles'.

As with writers of several other works which discuss meteorology, the author's primary motive is not to offer an explanation of meteorological phenomena, but to provide an account of the order and arrangement of the cosmos as a whole. *On the Cosmos* presents a unified view in which both the most ordinary and the most terrifying meteorological events are part of the orderly arrangement of the universe. These phenomena contribute to the preservation of that order and are all due to the same first cause. It is as part of this larger project that the author presents detailed catalogues of certain types of phenomena, specifically winds and earthquakes.

The overall plan of *On the Cosmos* may be summarized briefly. In the first chapter, the author commends the study of philosophy to Alexander, proposing to 'theologize' about the cosmos, and suggesting that the study of the greatest things is appropriate to the best leaders.[153] In chapter 2, he takes care to set out the terminology used in his discussion, defining words such as 'cosmos' (κόσμος), 'heaven' (οὐρανός), and 'stars' (ἄστρα), as he works through a description of the celestial region (that of the aetherial and divine nature), and the regions of Fire and Air. In the region of Fire, 'meteors and flames shoot across, and often planks and pits and comets, as they are called, stand motionless and then expire' (392b3–5). (In spite of the author's concern with providing definitions, there are inconsistencies. Later in the work, where he presents a fuller discussion of certain phenomena, planks, pits and comets (395b11–13) are located in the region of the Air.) It is in the region of the Air, 'which itself also has the power to change, and alters in every kind of way', that many other meteorological phenomena have their origin, including clouds, rain, snow, frost, hail, gales, whirlwinds, thunder, lightning, thunderbolts and storm-clouds.[154]

Following this outline of the regions of the elements of Fire and Air, the earth and sea (γῆ καὶ θάλασσα) are described in chapter 3. The terms chosen are striking, for they connote the actual physical land and oceans, rather than the elements of Water and Earth. The physicality of the terminology is borne out by the geographical and descriptive character of the chapter, which is presented as an account of the inhabited world. This geographical presentation foreshadows the author's repeated invocation of analogies with human social and political life throughout the work.[155]

The author explains his intention to discuss only briefly the most important phenomena of the inhabited world; it is in this section that meteorological phenomena are described in greatest detail.[156] He begins with a discussion of the two exhalations, one being dry and like smoke, the other damp and vaporous, and their products. Here, as in earlier sections, he takes pains to name specific phenomena and to define terminology. So, such terms as 'clear sky' (αἰθρία), 'downpour' (ὑετός), and 'snowstorm' (νιφετός) are specifically detailed.[157] Like Seneca, the author of *On the Cosmos* was interested in names, definitions and etymologies.

Wind is produced from the dry exhalation and is called *pneuma* ('breath'). The author notes that this has the same name as that which is present in plants and animals that brings life and generation, and states that no further discussion is needed. 'Wind' (ἄνεμος) and 'breeze' (αὔρα) are then differentiated, the first phenomenon deriving from the dry exhalation, the latter from the moist. The author gives a detailed discussion of the various winds, their names, directions and characteristics; the detail here is similar to that in the geographical description of the previous chapter.

Next, the author distinguishes between those phenomena of the air which are 'mere appearances' and those that are 'real'. The first group includes rainbows and streaks in the sky (ῥάβδοι), while the second includes lights, shooting stars and comets.[158] Because of the traditional attribution of the work to Aristotle, it is useful to compare the treatment of these phenomena in *On the Cosmos* to that in Aristotle's *Meteorology* and Theophrastus' *Meteorology/Metarsiology*. Aristotle regards rainbows, haloes, mock suns and rods as optical phenomena, due to reflection (*Meteorology* 372a17–21), but he does not make a similarly succinct distinction between 'mere appearances' and 'real' phenomena. While the author of *On the Cosmos* does not use precisely the same vocabulary as Aristotle, it is clear that he also regards rainbows and rods as optical phenomena.[159] Their shared view of haloes as due to reflection contrasts with that of Theophrastus. (In Theophrastus' *Meteorology/Metarsiology* there is no discussion of comets or of rainbows; it is possible that these phenomena were discussed in the original version, but no longer survive.) Theophrastus, unlike Aristotle and the author of *On the Cosmos*, regards haloes as materially constituted, and not as optical phenomena; he explains the lunar halo as occurring 'when the air becomes thick and is filled with vapor'.[160] Similarly, Seneca, like Theophrastus, regards haloes as due to dense air (*NQ* I.2.4). He treats rainbows, which he regards as due to reflection (*NQ* I.3.11), in the same book as haloes, which indicates that he does not organize his treatment according to the distinction made in *On the Cosmos*.

This distinction between meteorological phenomena which are 'mere appearances' (i.e. optical) and those which are 'real' (materially constituted) does not seem to be directly traceable either to Theophrastus or to Aristotle. The author concludes his brief discussion of both types of phenomena by noting that he has finished with the things of the air (τὰ μὲν τοίνυν ἀέρια τοιαῦτα); he then directs his attention to the location of the earth, before turning to those things which occur in the sea. His presentation makes it clear that he is focusing on the location of phenomena, rather than what might be regarded as the dominant elements (Air, Earth and Water) characteristic of each region within the terrestrial realm. He emphasizes the mingling of the various elements: 'to sum up, since the elements are mingled one with another, it is natural that phenomena in the air and land and sea should show these similarities, which involve destruction and generation for the individual parts of nature, but preserve the whole free from corruption and generation'

(396a28–33). The earth contains many sources of wind and fire, as well as water. There are many places on the earth where the surface has vents for the winds; these openings have various effects on people in the vicinity. Proximity to the wind vents may cause ecstatic inspiration, sickness, or even destruction; the author suggests (395b29) that the ability to prophesy in Delphi is related to the open vents for the winds.[161]

Winds cause earthquakes, which are catalogued in detail (395b26ff.).[162] The author argues that some phenomena of ocean waves are analogous to those produced in the earth by winds; these similarities are not surprising in the light of the co-mingling of the elements. Phenomena of the air, land and sea display similarities. But while these may involve destruction and generation (φθορὰς καὶ γενέσεις) for individual parts of nature, they nevertheless serve to preserve the whole. The teleological tone of this passage is developed as the treatise progresses.

The author describes the beauty of the order of the cosmos, asking:

> What particular detail could be compared to the arrangement of the heavens and the movement of the stars and the sun and moon, moving as they do from one age to another in the most accurate measures of time? What constancy could rival that maintained by the hours and seasons, the beautiful creators of all things, that bring summers and winters in due order, and days and nights to make up the number of a month or a year?

But it is not only the phenomena of the celestial, divine region which are ordered. Those of the terrestrial region, including the *meteōra*, also have their role in the larger cosmic design:

> even the unexpected changes are accomplished in due order – the winds of all kinds that dash together, thunderbolts falling from the heavens, and storms that violently burst out. Through these the moisture is squeezed out and the fire is dispersed by currents of air; in this way the whole is brought into harmony and so established.[163]

The author explains that even terrifying phenomena, which appear disorderly and unpredicted, are part of the whole and ultimately for good. And not only the spectacular but also the more ordinary *meteōra* contribute to the cosmic order:

All these things, it seems, happen for the good of the earth and give it preservation from age to age: for when it is shaken by an earthquake, there is an upsurge of the winds transfused within it, which find vent-holes through the chasms, as I have already said; when it is washed by rain it is cleansed of all noxious things; and when the breezes blow round about it the things below and above it are purified.[164]

In the author's cosmos, even everyday meteorological phenomena contribute towards the final cause, the teleological preservation of the universe. Furthermore, the *meteōra* share the same cause: the 'rains in due season, and winds, and falls of dew, and all the phenomena that occur in the atmosphere – all are the results of the first, original cause' (399a24f).[165]

But while the *meteōra* play a crucial role in the order of the cosmos, they occur only within the terrestrial region. It is characteristic of the earth to experience terrible weather and earthquakes (400a25ff.):

For violent earthquakes before now have torn up many parts of the earth, monstrous storms of rain have burst out and overwhelmed it, incursions and withdrawals of the waves have often made seas of dry land and dry land of seas; sometimes whole cities have been overturned by the violence of gales and typhoons; flaming fires from the heavens once burnt up the Eastern parts, they say, in the time of Phaëthon, and others gushed and spouted from the earth, in the West, as when the craters of Etna erupted and spread over the earth like a mountain-torrent.[166]

But these violent meteorological events are confined to the earth; the divine region is not affected. On this subject, the author cites the authority of the ancient poems, noting that the gods live outside of the realm of weather, according to Homer (citing *Odyssey* 6.42–5):

To Olympus, where they say the gods' dwelling stands
always safe; it is not shaken by winds, nor drenched
by showers of rain, nor does snow come near it; always
    unclouded
the air spreads out, and a white radiance lies upon it.[167]

Even if the gods reside outside of the realm affected by the *meteōra*, the author of *On the Cosmos* still regards the divine as closely linked to the production of these phenomena. (Notably, Lucretius translates this same passage from Homer, in his description of the abode of the gods in the prologue to Book 3, lines 18–24, of his poem.)

In a passage reminiscent of Seneca's discussion of the various names for Jupiter, the author of *On the Cosmos* lists the many names used for Zeus, explaining that 'though he is one, he has many names, according to the many effects he himself produces'. Many of these effects are meteorological: hence some of his names: 'God of Lightning and of Thunder, God of the Air and Aether, God of the Thunderbolt and the Rain – he takes his name from all these things'.[168] Citing the ancient authority of the Orphic books, the author (401a30–b8) quotes:

> Zeus is the first-born, Zeus is last, the lord of the
>   lightning; . . .
> Zeus is king, Zeus is the master of all, the lord of the
>   lightning.[169]

Here the Orphic books, like Seneca's Etruscans, credit Zeus/Jupiter as wielding lightning bolts. The author of *On the Cosmos*, like Seneca, shows a willingness to incorporate the traditional association of the gods with the *meteōra*, while pointing to larger philosophical issues related to causes:

> I think too that Necessity is nothing but another name for him [Zeus], as being a cause that cannot be defeated; and Destiny because he binds things together and moves without hindrance; Fate, because all things are finite and nothing in the world is infinite; Moira, from the division of things; Nemesis, from the allocation of a share to each; Adrasteia – a cause whose nature is to be inescapable; and Aisa – a cause that exists for ever.[170]

This list of names for Zeus, indicating various functions and attributes, signals a variety of philosophical influences. Various features of the work make it difficult to label it specifically as 'Peripatetic' or 'Stoic'; the author ends the work with a quotation from Plato's *Laws* on the role of divinity in the world.

The author of *On the Cosmos* clearly had a broad cosmological and theological motive in writing the work, and meteorological explanation was not his primary interest. He very likely relied on handbooks or epitomes for the section on meteorology. Although it is not clear

which handbooks those may have been, the account presented in
On the Cosmos indicates the sort of meteorological information and
explanation available to non-specialist readers.[171] Overall, the work
conveys a strong sense of the desire of those interested in cosmology
and theology to continue to engage with issues relating to the fears
and sense of disorderliness associated with the traditional gods and
their control of the *meteōra*, a desire shared by other philosophical
writers and poets of the period.

5

AN ENCYCLOPEDIC
APPROACH

In 1982, the Weather Channel was launched in the United States, a cable television station devoted to coverage of ongoing weather events and forecasts; in the year 2000, fifteen million tuned into the station at some point during the day. John Seabrook, in the *New Yorker* magazine, described the weather-tracking of 'weather junkies' as a 'national obsession';[1] 'weather junkies' track frightening meteorological events as they occur. Seabrook explains the public fascination with dire weather: 'Tornadoes, floods, hurricanes, thunderstorms, blizzards – who doesn't like to watch that stuff?'

There is evidence of fascination with meteorological dramas and disasters (including signs and portents) from antiquity to our own time. For example, Aristotle mentions the earthquake and associated tidal wave at Achaea twice in the *Meteorology* (343b2, 368b7f.); the first time he also refers to a comet seen at about the same time. In his poem, Lucretius uses vivid and evocative language for meteorological phenomena, as in his description of flames shooting out from the throat of Mount Aetna (6.641–6):

> It was no middling slaughter
> When the flame-storm rose and swept like a lord through
> the fields
> Of Sicily; neighboring peoples turned their eyes
> And saw the smoking temples of the sky
> Glitter with sparks; they trembled, for they feared
> What revolution Nature had in store.[2]

Seneca (*NQ* VI.27.1) provides a graphic account of the distress following the Campanian earthquake in 62 CE. As we have seen, many ancient meteorological works were written with the aim of dispelling fear of such natural disasters, by providing some understanding of their causes.

During the medieval period, continuing interest in meteorological phenomena is shown in works such as Isidore of Seville's (d. 636) encyclopedic *Etymologies*, in which he explains the nature of air and clouds (Book 13, chapter 7), thunder and lightning (chapter 8), and rainbows and rain (chapter 10); Isidore used Pliny (perhaps second hand) as a source.[3] Many commentaries on Aristotle's *Meteorology* were produced, in Hebrew, Arabic and Latin; it was the first work in the Aristotelian corpus to be translated into Hebrew.[4] In the early modern period, unusual and frightening meteorological events held a fascination for many; there are numerous examples of broadsheets published to alert people to meteorological events and their portents, including the appearance of comets and parhelia.[5]

The tradition of weather prognostication flourished in the past, as it does today, aided by various calendars and almanacs. A relatively large number of astrometeorological manuscripts survive from the fourteenth and fifteenth centuries; Stuart Jenks has described 139, which were probably used by astrologers, physicians and monks.[6] Vladimir Janković has argued that, in England during the seventeenth and eighteenth centuries, meteorology was concerned to explain events by means of a theory of sublunary elements, while weather was not subject to explanation, but only prediction.[7]

In mid-nineteenth century England, astrometeorologists vied with official, government-sponsored scientists to produce accurate forecasts. At the same time, almanacs were sold in the hundreds of thousands, and weather lore circulated through various means, including poems and weather ditties, such as 'sky red in the morning is a sailor's sure warning; sky red at night is the sailor's delight'.[8] In 1893, Richard Inwards, a fellow of both the Royal Meteorological Society and the Society of Antiquaries, published a collection of weather lore applicable to the climate of the British Isles; he gathered proverbs, sayings and rules from many sources, including a number of ancient authors. His collection aims to be 'a manual of outdoor weather wisdom seen from its traditional and popular side',[9] and gives a sense of the long tradition of interest in understanding and predicting meteorological phenomena. Homer is quoted once, Hesiod twice; Aratus and Virgil are referred to numerous times. Aristotle is cited as an authority on thirteen occasions, while the encyclopedic author Pliny the Elder is named fourteen times. Pliny was an unusual author on meteorological topics, for he dealt with both explanation and prediction in his *Natural History*; he also died as a direct result of a disastrous meteorological event.

Pamphlets and 'Wonderful news' about strange meteors in Stuart England

A Report from Abbington towne in Berkshire, being a relation of what harme Thunder and Lightning did on Thursday last upon the body of Humphrey Richardson, a rich miserable farmer. With Exhortation for England to repent. London: William Bowde, 1641.

A Strange Wonder or The Cities Amazement. Being a Relation occasioned by a wonderful and unusual Accident that happened in the River of Thames., Friday Feb, 4, 1641. London: printed for John Thomas, 1641.

A Strange and Wonderful Relation of the Miraculous Judgement of God in the late Thunder and Lightning. London: Bernard Alsop, n.d. [c. 1620–50]

Charles Hammond, *A Warming-peece [sic] for England. By that sad and fearfull Example that happened to Men, Women and Children, all sorts of cattle and fowle; by Stormes, Tempests, Hail-Stones, Lightning and Thunder, June 25 1652.* London, 1652.

Mirabilis Annus (Eniautos Terastios) or the year of Prodigies and Wonders, being a faithful and impartial Collection of several Signs that have been seen in the Heavens, in the Earth, and in the Waters ... all which happened within the space of year last past, and are now made publick for a reasonable Warning to the People of these Kingdoms speedily to repent and turn to the Lord ... London, 1661

This Winters Wonders: or a True Relation of a Calamitous Accident at Bennenden in the Country of Kent; how the Church and several Houses were destroyed by thunder and Lightning, on the 29th of December last, being Sabbath day. London, 1671.

Strange and Terrible News from Sea. Or, A True Relation of a Most Wonderful Violent Tempest of Lighting and Thunder on Friday, the 18th of this Instant Jan. 1678. Westsmithfield: A. P. and T. H. for John Clarke, 1678. 6 pp

Memento to the World: Or An Historical Collection of Divers Wonderful Comets and Prodigious Signs in Heaven, that have been seen, some time before Birth of Christ, and many since that time in divers Countries, with their wonderful and dreadful Effects. London: T. Haly, 1680.

T. T., *Modes Observations on the Present Extraordinary Frost: Containing I A Brief Description thereof, and its Natural Celestial Causes inquired into. II An Account of the most eminent Frosts that have happened [in the past]. III Philosophical presages of what may be feared now to ensue [scarcity, sickness, pestilence]. IV The cries of the poor, and an easy way proposed how there may be 20 000 pounds a week raised for their relief ...* London: George Larkin, 1684

Figure 5.1 Pamphlets and 'wonderful news' about strange meteors in Stuart England.

Source: V. Janković (2000) *Reading the Skies: A Cultural History of English Weather, 1650–1820*, Chicago: University of Chicago Press, 38. Copyright © Vladimir Janković 2000, reprinted with permission.

Pliny the Elder

The eruption of the volcano Vesuvius in 79 CE has its place in the public imagination as one of the most memorable natural events of antiquity.[10] One of those who perished during the eruption that buried the city of Pompeii was Gaius Plinius Secundus (23/4–79 CE), usually known as Pliny the Elder. On the fatal day in 79, Pliny led a rescue operation to the disaster area. His nephew, Pliny the Younger, in a letter written some twenty-five years later to the historian Tacitus, described his uncle's activities on that day, and says that the first sign of the eruption was a 'cloud':

> My uncle was stationed at Misenum in active command of the fleet . . . in the early afternoon, my mother drew to his attention a cloud of unusual size and appearance. He had been out in the sun, had taken a cold bath, and lunched while lying down, and was then working at his books. He called for his shoes and climbed up to a place that would give him the best view of the phenomenon. It was not clear at that distance from which mountain the cloud was rising (it was afterward known to be Vesuvius). . . . My uncle's scholarly acumen saw at once that it was important enough for a closer inspection, and he ordered a boat to be made ready. . . .
>
> As he was leaving the house he was handed a message from Rectina, wife of Tascus, whose house was at the foot of the mountain, so that escape was impossible except by boat. She was terrified by the danger threatening her and implored him to rescue her from her fate. He changed his plans, and what he had begun in a spirit of inquiry he completed as a hero. He gave orders for the warships to be launched and went on board himself with the intention of bringing help to many more people besides Rectina, for this lovely stretch of coast was thickly populated.
>
> He hurried to the place which everyone else was hastily leaving, steering his course straight for the danger zone. He was entirely fearless, describing each new movement and phase of the portent to be noted down exactly as he observed them. Ashes were already falling hotter and thicker as the ships drew near, followed by bits of pumice and blackened stones, charred and cracked by the flames: then suddenly they were in shallow water and the shore was blocked by the debris from the mountain.[11]

Pliny the Elder paid for his curiosity; he died from inhaling the toxic fumes of Vesuvius.

While Pliny is well known for having perished in the disaster, his reputation is founded on his encyclopedic thirty-seven-book work, the *Natural History* (*Naturalis Historia*), which aimed to cover the full range of contemporary knowledge. The first book lists the contents of the whole and the authorities consulted; the topics covered a vast range, from the question whether the world is finite (Book 1) to notable fish-ponds (Book 9), the craftsmanship of birds in nest-making (Book 10), and when duels of gladiators were first painted and exhibited (Book 35). At certain points in the *Natural History* Pliny's aim seems to be to present a summary not only of contemporary knowledge, but also of different points of view. The dual themes of curiosity and duty play throughout the work; it is filled with lists of information and wonders, with a sprinkling of moral exhortation.

Pliny's discussions of the prediction and explanation of meteorological phenomena are largely presented in separate sections of the *Natural History*. His account of their causes occupies relatively brief sections, but the discussions are importantly located, in the middle (conceptually) of his wider account of his understanding of the nature and workings of the world. He describes and discusses the causes of what he refers to as the phenomena of the air in Book 2, as part of his larger examination of the world. Agriculture, arboriculture and gardening are discussed in Books 17–19; he treats weather prediction largely within the context of farming activities in Book 18.

As a member of the privileged equestrian order, Pliny, like many of his class, prided himself on his ability and acumen in practical matters. He presents himself as a man of the earth, greatly interested in agriculture. In Book 18, he states that agriculture is the most fundamental human activity. For him, farming is dependent on astronomy, because of its usefulness in determining the seasons, establishing an agricultural calendar and predicting weather.

Chapter 2 showed that astrometeorological *parapēgmata* and almanacs enjoyed a long history. Though few examples of these stone *parapēgmata* survive, it is clear that they formed part of a long-lived practice, adopted within Roman culture as well as Greek. A fragment of a Latin *parapēgma,* found at Puteoli (another Campanian town, 12 km north of Naples),[12] is inscribed with the Roman numeral 'XII' and correlates the setting of Delphinus with a storm; individual peg holes are provided for both the date (XII) and the weather prediction.[13] Examples of solar calendars known as the *Menologia Rustica* also survive. In these, the sun's position in the

Figure 5.2 Parapēgma fragment from Puteoli (modern Pozzouli).

Source: Museo Archeologico Nazionale, Naples.
Note: See Degrassi (1963) 310; Mingazzini (1928).

Figure 5.3 Drawing after the *Menologium Rusticum Colotianum*, attributed to Jean Matal, late sixteenth century.

Source: © Copyright The British Museum.

zodiac is only roughly given, in order to associate each zodiacal sign with a particular month; the *Menologia* (or Farmers' Calendars) do not give precise stellar phases.[14] (There are Roman examples of other sorts of *parapēgmata*, for example some which are strictly calendrical, in which the peg is moved daily to keep track of the date, but no astronomical or meteorological information is given.[15])

Chapter 2 also showed that the production and promulgation of astrometeorological almanacs were not solely the province of astronomers; several notable poets, including Aratus, also played a role. Further, a number of Roman agricultural writers, including Varro (116–27 BCE; *On Agriculture* 1.27–37) and Columella (fl. 50 CE; *On Agriculture* 11.2) produced almanacs at least partially defined by astronomical events; these served as farmers' almanacs.[16] The ancient agricultural writers describe farmers using astronomical observations to predict the seasons and the weather. In Book 1 of the *Natural History*, Pliny states that he has used the works of both Varro and Columella; among the poets, he refers to Hesiod and Virgil as authorities for his own work.

In the *Georgics* Virgil shows the importance of being familiar with the seasons and the weather, and of using astronomical knowledge (1.50, 1.252–8), explaining that:

> Thus can we forecast weather though the sky
> Be doubtful, thus the time to reap or sow,
> When best to impel the treacherous sea with oars
> And launch armadas, when to fell the pine-tree.
> For not in vain we watch the constellations,
> Their risings and their settings, not in vain
> The fourfold seasons of the balanced year.[17]

Virgil himself served as 'role model' for Pliny, who cited the poet in the presentation of his farmers' calendar. Pliny notes (*NH* 18.56.206), clearly referring to the passage above, that 'Virgil enjoins first before all else to learn the winds and the habits of the stars, and to observe them just in the same way as they are observed for navigation.'[18]

Pliny presents his farmers' almanac (*NH* 18.60.224–74.320), which includes star phases and associated weather predictions, as well as agricultural advice. It is lengthy and detailed, and he acknowledges various authorities, including Virgil, Cato (234–149 BCE) and Julius Caesar (100–44 BCE), as providing specific information and recommendations.[19] His coverage begins with the proper date

for sowing crops, which he says 'needs very careful consideration' (*NH* 18.56.201). He notes that 'Hesiod, the leader of mankind in impart-ing agricultural instruction, gave only one date for sowing, to begin at the setting of the Pleiads', but Pliny points out that this date is determined by the region in which Hesiod lived, Greek Boeotia.[20]

Here is a passage from Pliny's almanac (*NH* 18.65.237), in which he offers specific dates, stellar risings and settings, and weather predictions:

> Between the period of west wind and the spring equinox, February 16 for Caesar marks three days of changeable weather, as also does February 22 by the appearance of the swallow and on the next day the rising of Arcturus in the evening, and the same on March 5 – Caesar noticed that bad weather took place at the rising of the Crab, but the major-ity of the authorities put it at the setting of the Vintager.[21]

He goes on to explain (*NH* 18.65.238) that:

> This space of time is an extremely busy period for farmers and especially toilsome, and it is one as to which they are particularly liable to go wrong – the fact being that they are not summoned to their tasks on the day on which the west wind ought to blow but on which it actually does begin to blow. This must be watched for with sharp attention, and is a signal possessed by a day in that month that is observable without any deception or doubt whatever, if one gives close attention.[22]

Pliny explains that astronomical knowledge can be useful in agricul-ture, but he emphasizes (*NH* 18.75.321) that he is primarily interested in the practical value of general rules, rather than trying to assign particular operations to specific days. He stresses the need for careful observation by the farmer himself.

Pliny was aware of his dependence on the tradition of astro-meteorology, but he was able to question its reliability. Throughout the *Natural History*, he relies on information and ideas taken from those whom he names in the first book, in the outline of the contents. But he is, at times, also aware of the problems inherent in relying on the work of others – this in spite of his credulity regarding 'tall tales' and wonders.[23] So, for example, in Book 18 he comments on the difficulties caused by various authors having made observations at different locations, stating, at some length, that observations and predictions cannot be generalized from place to place.[24] He notes that the different localities at which individuals worked would produce differences in what was observed by each. But he also noted that astronomers working in the same region did not always agree. This is significant in that the information attributed to the astronomers cited as authorities in the *parapēgmata* does not always agree. He gives an example:

> the morning setting of the Pleiads is given by Hesiod – for there is extant an astronomical work that bears his name also – as taking place at the close of the autumnal equinox, whereas Thales puts it on the 25th day after the equinox, Anaximander on the 30th, Euctemon on the 44th, and Eudoxus on the 48th.[25]

(It is worth noting that Pliny names Hesiod, subtly indicating that he is working in the same tradition, providing practical advice while being concerned with wider ethical issues.)

As part of the larger treatment of the timing of agricultural operations and methods of weather forecasting in Book 18, Pliny outlines how a wind-rose may be drawn on the ground using simple astronomical principles (18.76.326ff.), or carved on a wood block (18.77.331–2). The wind-rose or compass may be used to determine wind direction, enabling the wind to be identified; several ancient Roman marble wind-roses (anemoscopes) survive.[26] The description of the construction of a wind-rose occurs in his discussion of individual winds (18.77.333–9, which he identifies by their Greek names), their relations to and effect on farm activities. In this section he mentions the views of Aristotle (whom he calls a man of immense acuteness; *vir immensae subtilitatis*) and Cato.

In Book 18, Pliny presents his farmers' almanac, and provides detailed recommendations for farm activities. At the end of it, he lists signs useful for predicting the weather, including those given by the sun, moon, stars, meteorological events (such as thunder and lightning), fire (such as flickering or sparking), the sea, animals and birds. His treatment in Book 18 is, for the most part, limited to providing lists of useful recommendations for landowners and their workers. (He does discuss the difficulty of properly defining the year, astronomically.) In using the almanac, Pliny cautions against going too closely 'by the book': the man of the land must make his own observations; the various weather signs he provides show the sorts of

Figure 5.4 Marble wind-rose (anemoscope), *c.* 200 CE, found south of Porta
          Capena, near the Via Appia in Rome.

Source: Biblioteca e Musei Oliveriana, Pesaro.
Note: See also Figure 4.2, for another view.

indications to be looked for. The proper farmer should not be too
much in thrall to astronomers and authorities; he should try to use
their wisdom, when appropriate, but not be content to stick to 'rules';
rather, he must look out (literally) for himself.

Exceptionally for authors writing on prediction, Pliny also, in
Book 2 of the *Natural History*, gives detailed consideration to the
causes of meteorological phenomena. In his dual aim of explaining
and predicting meteorological phenomena, Pliny is unusual. In his
presentation of his farmers' almanac and listing of weather signs,
Pliny makes it clear that he is building on the work of others.
Similarly, in his meteorological explanations Pliny discusses the

observations and ideas of predecessors; for example, Aristotle is mentioned (approvingly) as an authority several times in Book 2. Pliny's
explanations of meteorological phenomena show the influence both
of Aristotle and of Theophrastus. Exhalations play a role in causing
weather, as do the motions of the celestial bodies. Like Aristotle,
Pliny acknowledges that the causes of meteorological phenomena are
not entirely understood; and like Theophrastus, he finds multiple
explanations acceptable.

For Pliny, the winds are pre-eminent and are the cause of most
other phenomena of the air. Before launching into his own detailed
description of the winds, he provides (*NH* 2.45.17–18) a brief, somewhat moralizing, account of previous work on the subject. He
marvels at the efforts of over twenty Greek authors, working under
adverse conditions, to study the winds, explaining that:

> This makes me all the more surprised that, although when
> the world was at variance, and split up into kingdoms, that
> is, sundered limb from limb, so many people devoted them-
> selves to these abstruse researches, especially when wars
> surrounded them and hosts were untrustworthy, and also
> when rumours of pirates, the foes of all mankind, terrified
> intending travellers – so that now-a-days a person may learn
> some facts about his own region from the notebooks of
> people who have never been there more truly than from the
> knowledge of the natives.

He bemoans the lack of interest of contemporary Romans in research
into the subject, complaining that:

> now in these glad times of peace under an emperor who so
> delights in productions of literature and science, no addition
> whatever is being made to knowledge by means of original
> research, and in fact even the discoveries of our predecessors
> are not being thoroughly studied.

He contrasts the scholars of a bygone age who pursued the study of
the winds for mainly intellectual reasons, with people of his own day
who are interested only in getting rich:

> The rewards were not greater when the ample successes were
> spread out over many students, and in fact the majority of
> these made the discoveries in question with no other reward

at all save the consciousness of benefiting posterity. Age has overtaken the characters of mankind, not their revenues, and now that every sea has been opened up and every coast offers a hospitable landing, an immense multitude goes on voyages – but their object is profit not knowledge; and in their blind engrossment with avarice they do not reflect that knowledge is a more reliable means even of making profit.[27]

Acknowledging the utility of knowledge of the winds, for navigation and trade, Pliny explains that his own account of them is more detailed than it otherwise might have been, for the benefit of those who go to sea. He wants his discussion to have practical use. Pliny explains that

> there is a great difference between a gust of air and a wind. The latter, regular and blowing steadily, and felt not by some particular tract only but by whole countries, and not being breezes nor tempests but winds – even their name being a masculine word – whether they are caused by the continuous motion of the world and the impact of the stars travelling in the opposite direction or whether wind is the famous 'breath' that generates the universe by fluctuating to and fro as in a sort of womb, or air whipped by the irregular impact of the planets and the non-uniform emission of their rays, or whether they issue forth from these nearer stars which are their own or fall from those stars which are fixed in the heavens – it is manifest that the winds too obey a law of nature [*legem naturae*] that is not unknown, even if not yet fully known.[28]

Pointing to the limits of contemporary knowledge, Pliny distinguishes between these natural law-abiding winds and gusts of air; the latter may be produced by a dry and parched exhalation from the earth, or when bodies of water exude a vapour which is neither mist nor cloud, or by the force of the sun driving the air, or by many possible causes.[29]

He provides a brief history of the designation of the winds: 'The ancients noticed four winds in all, corresponding to the four quarters of the world'; he remarks that 'even Homer mentions no more', implying that he takes Homer as a great authority. But Pliny is critical of what he called this 'dull-witted system', which he believed was not improved by successors who named an additional eight winds;

this designation he judged to be 'too subtle and meticulous'. Eventually a compromise was devised, by adding four from the long list to the original four in the short list, so that two winds in each of the four quarters of the heaven were distinguished. Pliny character- izes the winds, naming each and saying from whence they blow, and relates the seasons in which each blows, as well as their effects.[30] Pliny concludes his review by noting that Eudoxus thought that all sorts of bad weather, not only winds, recurred in four-yearly cycles, begin- ning with the rising of Sirius; this remark reinforces Pliny's own view about the regularity of the winds, and of nature generally.[31]

Pliny distinguished between winds, which exhibit some degree of regularity, and sudden gusts of air, caused by exhalations from earth. The latter sometimes interact with clouds, producing cloudbursts, typhoons, the *prēstēr*, pillars, tornadoes, whirlwinds, columns and waterspouts. He discusses various phenomena of the air, describing rain, storms, the influence of the heavenly bodies on weather, clouds, winds, whirlwinds, thunderbolts, thunder, lightning, the relation- ship of thunderbolts to prayer, rainbows, hail, snow and frost. These are mostly described briefly, although in some cases (e.g. thunder- bolts) the associated portents are also related.[32] Into this list, he interjects accounts of other phenomena, recounting specific examples of various extraordinary phenomena of the air, including the raining of milk, of blood, of flesh, of iron in the shape of sponges, of wool and of baked bricks;[33] the appearance and sound of armies in the sky on several occasions are also noted.[34] Pliny reports these occurrences matter-of-factly, providing details of the time and place in which they were witnessed.[35] Generally, no explanation is offered, although he accounts for the appearance of armies marching through the sky as being due to the sky itself catching fire when clouds have been set on fire by a large flame. He also discusses stones falling from the sky, noting that one such stone is worshipped in Abydos, but he dismisses the prophecies of Anaxagoras regarding such stones.[36] As part of this larger discussion of phenomena, Pliny states clearly (*NH* 2.60.150) that he does not regard rainbows as portents. This section of the *Natural History* seems, to the modern reader, an amazing blend of the credulous and the sceptical, as Pliny reports sightings of wool rain- ing from the sky but condemns the divining powers of Anaxagoras, rejecting the view that the sun is a stone because such a belief would undermine our understanding of the physical universe.

Throughout his discussion, Pliny presents a dramatic picture of the phenomena of the region of the air, noting that 'from it come clouds, thunder-claps and also thunder-bolts, hail, frost, rain, storms and whirlwinds; from it come most of mortals' misfortunes, and the warfare between the elements of nature'. It is like a battle scene, with the winds committing acts of pillaging:

> Rain falls, clouds rise, rivers dry up, hailstorms sweep down; rays scorch, and impinging from every side on the earth in the middle of the world, then are broken and recoil and carry with them the moisture they have drunk up. Steam falls from on high and again returns on high. Empty winds sweep down, and then go back again with their plunder. . . . Thus as nature swings to and fro like a kind of sling, discord is kindled by the velocity of the world's motion.[37]

The account must be meant to inspire wonder and awe.

Yet Pliny's explanation incorporates themes familiar from the accounts given by natural philosophers. Pliny recognizes the influence of celestial bodies as an important cause of meteorological phenomena. And when he briefly (2.39.105) touches on storms and rain, he explains that, while they clearly have some regular causes, others are accidental or not fully explained. In his view, the motions of the celestial bodies, the fixed stars as well as the planets (which include the sun and moon), influence the seasons and the weather. Asking 'who can doubt that summer and winter and the yearly vicissitudes observed in the season are caused by the motion of the heavenly bodies?', he answers that:

> as the nature of the sun is understood to control the year's seasons, so each of the other stars also has a force of its own that creates effects corresponding to its particular nature. Some are productive of moisture dissolved into liquid, others of moisture hardened into frost or coagulated into snow or frozen into hail, others of a blast of air, others of warmth or heat, others of dew, others of cold.

He explicitly states that both the planets and the fixed stars influence weather. He points to the widespread recognition that the increase in the heat of the sun coincides with the rising of the Dog-star.[38]

But he is careful not to exclude other causes. He notes that 'it is certain that the earth exhales a damp mist and at other times a smoky one due to vapour, and that clouds are formed out of moisture rising to a height or air condensed into moisture'.[39] He goes on to present several possible causes of storms. It is possible for the fires of stars to fall into clouds, producing steam when they reach the cloud, and so generating storms. In his description of this steam, Pliny uses the analogy of a red-hot iron plunged into water. He uses another homely analogy to explain thunder: it is also possible for breath emerging from the earth, when pressed down by the counter-impact of the stars, to be checked by a cloud and so cause thunder, nature choking down the sound while the struggle goes on but the crash sounding when the breath bursts out, as when a skin is stretched by being blown into.[40] Similarly, the earthy exhalation may be struck by the impact of clouds, like two stones striking each other, and causing heat-lightning to spark. Offering these possible modes of causation, Pliny explains that these are all chance causes (*sed haec omnia esse fortuita*), which though they strike mountains and seas are without real effect, for they obey no rational nature. These 'chance' causes of thunder and lightning contrast with the thunderbolts described earlier in Book 2, those caused by the stars, by fixed causes, and which are prophetic.[41] In his discussion of the celestial region, Pliny mentions the view that thunderbolts originate in the three upper planets, particularly Jupiter. He explains that this is the source of the myth that thunderbolts are javelins hurled by the god Jupiter.[42]

Thunderbolts are also related to the earth's exhalation, in Pliny's view. He explains that Nature regulates the thunderbolt and lightning so that they occur in tandem. He notes that no one hit by lightning sees the flash or hears the thunder in advance, and considers that flashes occurring on the left are lucky. He notes that Tuscan writers thought that thunderbolts were sent by nine gods, but that only two of these, including Jupiter, were retained by the Romans. He says that records exist indicating that thunderbolts were brought about in response to prayer, and relates several examples. He notes, however, that:

> It takes a bold man to believe that Nature obeys the behests of ritual, and equally it takes a dull man to deny that ritual has beneficent powers, when knowledge has made such progress even in the interpretation of thunderbolts that it can prophecy that others will come on a fixed day, and whether they will destroy a previous one or other previous ones that are concealed: this progress has been made by public and private proofs in both fields.

He acknowledges the difficulty of taking a firm line here, remarking that 'such indications are certain in some cases but doubtful in others,

and approved to some persons but in the view of others to be con-
demned'.[43] Pliny then details the Tuscan methods for observing the
portents of thunderbolts, and follows it with a description of the
effects of lightning on humans, animals and plants.[44]

Rainbows are discounted as portents of anything, for they do not
even reliably predict rain or good weather. He says that it is clear that
they are caused by a ray of sun striking a hollow cloud, being repelled
and reflected back to the sun; the colours of rainbows are due to a
mixture of clouds, fire and air. Rainbows are only seen facing to the
sun, are always semicircular in shape, and not seen at night. Pliny
does, somewhat grudgingly, acknowledge that Aristotle states that
rainbows can sometimes be seen at night.[45]

Pliny almost races through his account of other phenomena,
including hail, snow and hoar frost, suggesting that these are well
known: 'hail is produced from frozen rain and snow from the same
fluid less solidly condensed, but hoar frost from cold dew; . . . snow
falls during winter but not hail, and hail itself falls more often in the
daytime than at night, and melts much faster than snow'.[46] His treatment of these phenomena is brief; he wants now to turn to his next
subject, Mother Earth. Pliny extols this region of the natural world
on which 'we have bestowed the venerable title of mother'.[47]

It is here, in the sections on the earth, that Pliny treats earthquakes
(2.81.191ff.). He begins with the theory credited to the Babylonians,
according to which earthquakes and ground fissures are caused by
the very same three stars (wandering stars, or planets), Saturn, Jupiter
and Mars, which also cause thunderbolts. Pliny himself believes
that the cause of earthquakes is the wind, 'for tremors never occur
except when the sea is calm and the sky so still that birds are unable
to soar because all the breath that carries them has been withdrawn;
and never except after wind, doubtless because then the blast has
been shut up in the veins and hidden hollows of the sky'. And in
his view, 'a trembling in the earth is no different from a thunderclap
in a cloud, and a fissure is no different from when an imprisoned
current of air by struggling and striving to go forth to freedom causes
a flash of lightning to burst out'.[48] He describes the various ways in
which earthquakes can occur, as well as their consequences. Like
lightning, earthquakes are more frequent in autumn and spring;
signs of impending earthquakes include sudden waves on the sea
unaccompanied by wind. There are great differences in the sorts of
movement which take place during earthquakes, but they stop once
the wind has found an escape.[49] Pliny concludes his account by noting
specific examples of earthquakes, some of which were portents of

greater disasters. Before his first-hand and fatal experience at Vesuvius, Pliny does not seem to have been particularly interested in volcanoes.

Pliny's treatment of meteorology is unusual in that it combines the desire to explain meteorological phenomena with the aim of prediction. Like many other ancient authors, he shares the goal of dispelling fear of natural phenomena. Yet, at times, (as Lucretius had done) he presents a vividly dramatic account. He also reports accounts of other wondrous events, which he presumably regards as being related to meteorology. His discussion of meteorology itself is riddled with contradictions: he bemoans the lack of interest in understanding the winds, while vaunting the practical utility of such knowledge. He advocates autopsy (seeing things for oneself), while recognizing the authority of ancient poets, astronomers, philosophers and agricultural authors; he generally takes portents seriously, while dismissing some (e.g. rainbows). Renowned in later times for his credulity, he is also at times blatantly sceptical.

Unusually for ancient authors, Pliny is:

> an author who never had to be 'rediscovered' since he was never lost. For many later readers, Pliny's account of ancient meteorological explanation and prediction would have been the only source of ancient approaches to these topics. There is evidence for at least some knowledge of Pliny during virtually every century from his own time down to the present.[50]

As Charles G. Nauert Jr explains, 'the only serious defect which ancient scholars found in the [*Natural History*] was a consequence of one of its principal virtues: it was exceedingly long. Hence to make a complete copy . . . was a laborious and costly task.'[51] It is worth noting the continuing interest in Pliny's *Natural History*, through the middle ages and early modern period. As Nauert points out, the topical organization of the work allowed readers and copyists to focus on those sections in which they were particularly interested. During the medieval period, there was a tendency to excerpt portions of the *Natural History*; for example, the sixth-century Johannes Lydus incorporated sections into his own work on portents.

At least 200 medieval manuscripts (in some cases partial copies) of the *Natural History* survive, showing the accessibility of the work throughout the period. The *Natural History* also attracted leading humanist scholars during the fourteenth, fifteenth and sixteenth centuries, even those who criticized Pliny's reliability as an authority.[52] In spite of questions about his work, Pliny continued to serve as an important source of information in many areas, including natural history; he was also regarded as a fund of curious anecdotes.[53] The *Natural History* was first printed in Venice in 1469; there were fifteen printed editions through 1500. Nicolas Jenson produced the first translation (into Italian) in 1476; the first dated French translation was published in 1544, and the first English translation in 1601 (by Philemon Holland).[54] During the sixteenth and seventeenth centuries, there was a proliferation of editions and commentaries, showing that Pliny was regarded as a useful source and authority on the natural world. Until at least the end of the eighteenth century, English agricultural writers valued the empirical evidence gathered in ancient prognostic weather texts; Pliny was among the authors of the texts most highly prized, along with Theophrastus (thought to be the author of *On Weather Signs*), Aratus and Virgil.[55]

In writing his *Natural History*, Pliny the Elder aimed to provide a summary of knowledge in his time. He presents a rich and somewhat complicated picture of meteorological explanation and prediction in antiquity; he deliberately incorporated Greek as well as Roman sources, information and ideas. His account is neither exhaustive nor complete, but it presents the range of ancient approaches, questions and concerns. Like many of his predecessors (including Aristotle), he acknowledges that the workings of nature may not be fully comprehended. Like many of those who sought to explain the *meteōra*, he is willing to suggest a number of possible causes. While he recognizes limits to prediction, and cautions against a blind allegiance to astronomers, he advocates the diligent use of observation to assay signs of weather. He is confident that rational causes operate in nature, even if they are not entirely known to us.

This assumption of the accessibility of the phenomena to rational understanding is a key characteristic of Greek and Roman meteorological explanations. The ancient texts show a strong desire to explain and predict ordinary phenomena (like rain), mingled with the urge to explain, if not actually predict, unusual, extraordinary and even wondrous events, while alleviating fear of such events. Remarkable occurrences (e.g. being struck by lightning, or witnessing showers of wool) are deliberately described as natural, not supernatural.[56] But, while there is the strongly held view that meteorological phenomena can be explained rationally, there is no assumption that the phenomena can be completely explained; many ancient attempts to explain the *meteōra* come with warning labels attached, cautioning readers that the explanation is incomplete. Indeed, the difficulty of dealing with meteorological phenomena, whether predicting or explaining them, is a recurrent theme.

The idea of underlying interconnections present in the universe is another important theme in the ancient meteorological texts. Various links between different parts of the natural world are depicted in the meteorological literature; what their precise nature might be is not always made explicit. For some authors, a 'link' may be an apt and convenient analogy offered as part of an explanation, without the intention of arguing for necessary causal connections. In other instances, links are meant to represent real connections within the cosmos. So, for example, the presumption of some link between the heavens and the earth underpinned some of the predictive techniques which were employed, collected, shared and transmitted over many centuries. But the relationship between cause and effect, between ability to explain and success in prediction is not made clear, and is usually not addressed.

From surviving ancient texts, it appears that few of those authors interested in explaining the *meteōra* aimed to make predictions; similarly, those authors who collected weather signs or correlated astronomical and meteorological phenomena did not, for the most part, attempt to explain them. But both types of texts, those dealing with weather prediction and those focused on meteorological explanation, show a deliberate engagement with the work of predecessors. The ancient texts on meteorology point to an area of the study of nature in which practitioners and authors built upon the work of others. In some cases the ideas and theories (and occasionally observations) of earlier workers are rejected; but even as ideas are criticized, they are discussed and examined and so incorporated into the literature, and given a means of survival in the intellectual marketplace.

The use of observations, explanations and predictive 'rules' made even in the distant past is evident throughout the ancient meteorological literature, in both poetic and prose texts. In some later works, the authority of the most ancient poets, Homer and Hesiod, is invoked, sometimes by name, sometimes not. Other, unnamed (and unnameable) sources are pointed to in the form of proverbs and adages. So, for example, Theophrastus, in *On Winds*, quotes approvingly from unattributed sayings: 'a north-wind rising in the night, never sees the third day's light', 'a frost hoar, southern winds roar'.[57] This valuing and preservation of culturally shared 'received wisdom' is evident throughout the Greek and Latin meteorological literature.

Of course, respect for weather lore is not confined to the ancient period and carries on well into modern times. In the eighteenth and nineteenth centuries in England, for example, ancient weather lore was taken to preserve important empirical facts.[58] In 1918, Vilhelm Bjerknes, leader of the 'Bergen' school of meteorology and one of the originators of the concept of the weather front, travelled around the Norwegian coast, collecting weather folklore from farmers, lighthouse keepers, fishermen and sailors.[59] Even at the beginning of the twenty-first century, collections of traditional weather lore are published annually, in the cheap format of 'farmers' almanacs', and are endorsed by readers as 'very accurate' and 'usually on the mark'.[60]

The reliance on the work of named and unnamed predecessors, in shaping both predictions and explanations, is a key characteristic of Greek and Roman meteorology. Coping with meteorological events was a sort of shared 'project', motivated by a number of concerns, including the desire to understand frightening and dangerous phenomena. The efforts of many people were incorporated into the ancient texts. Ancient authors, participants in an extended intellectual community, used and built upon the ideas, observations and prognostications of previous contributors to projects of predicting and explaining meteorological phenomena. The texts show the authors and their predecessors working together to come to terms with meteorological events.

NOTES

1 ANCIENT METEOROLOGY IN GREECE AND ROME: AN INTRODUCTION

1 Plato *Phaedrus* 270a; cf. *Cratylus* 396c.
2 For example, Aristotle *Meteorology* 351b35f. cites Homer (*Iliad* 9.381; cf. *Odyssey* 4.83–5, 229ff., 14.245ff., 295) when he discusses evidence of the effect of silting.
3 In the *Iliad* (9.5–8) the grief-stricken state of the Achaians is described by comparison to the winds:

> As two winds rise to shake the sea where the fish swarm,
>   Boreas and
> Zephyros, north wind and west, that blow from Thraceward,
> suddenly descending, and the darkened water is gathered
> to crests, and far across the salt water scatters the seaweed;
> so the heart in the breast of each Achaian was troubled.

Trans. Lattimore (1951) 198. See Willcock (1996); see also Cronin (2001) on weather lore as a source of Homeric imagery.
4 I have here given the line numbers for the Lattimore translations of the *Iliad* and the *Odyssey*.
5 Trans. Lattimore (1965) 98.
6 Hesiod *Theogony* 690–5, trans. Lombardo (1993) 80.
7 Hesiod *Theogony* 820–85.
8 Guthrie (1962) 450 n. 2 thought that neither Plato nor Aristotle 'can be supposed to have been very serious in this'. Cf. Ross (1924) 1: 130, on *Metaphysics* A 983b27f., who remarks that Plato 'jestingly suggests that Heraclitus and his predecessors derived their philosophy from Homer, Hesiod and Orpheus', and adds that Aristotle admits that 'the suggestion has no great historical value'.
9 See Lamberton and Keaney (1992), especially Lamberton's 'Introduction', particularly xvi, Richardson (1992) and Long (1992).
10 It is interesting to note that another Hesiodic work, the *Astronomy* (of which only fragments survive), may also have included a sort of calendar. Evelyn-White (1914) xix–xx suggests that it 'gave some account of the

principal constellations, their dates of rising and setting, and the legends connected with them, and probably showed how these influenced human affairs or might be used as guides'. See also West (1978) 23. The poem and the tradition will be discussed in some detail in the next chapter.

11 It is not clear that the last two, summer (θέρος) and late summer/early autumn (ὀπώρη), are to be distinguished (they are mentioned together at *Od.* 11.192 and 14.384–5.

12 As Bickerman (1980) 27 noted, 'Homer is reticent about any calendar . . . he mentions no month names.' A Homeric year seems to be seasonal: the year goes wheeling around and the same seasons return (*Od.* 9.294; cf. Hesiod *Theogony* 58; *Works and Days* 561).

13 See West (1978) 376–81, 'Excursus II: Time-reckoning', esp. 379–80 on stellar risings and settings.

14 *Airs, Waters, Places* 1.

15 See Hadzsits (1935).

16 Gilbert (1907); Kahn (1960) 99 n. 3.

17 Kahn (1960) 99, 109.

18 See Mansfeld (1992b) 320–2.

19 See Taub (1993) 138–9.

20 See Sambursky (1956b) for an overview.

21 Isager and Skydsgaard (1992) 10–11. See also Sallares (1991) 390f.

   Several ancient authors discuss long-term change. For example, in the *Meteorology* (351a19ff.), Aristotle considered the silting of rivers (cf. Sallares (1991) 391f.); in *On Winds* 13 Theophrastus mentions the view that winters in the past were more severe.

2 PREDICTION AND THE ROLE OF TRADITION: ALMANACS AND SIGNS, *PARAPĒGMATA* AND POEMS

1 The astrometeorological texts known as *parapēgmata* will be discussed below. Lehoux (2000) 90–2 and 117 has argued that, although such texts are sometimes referred to as 'calendars' or 'star-calendars', they are actually 'extra-calendrical tools for tracking phenomena which are not directly linked to one's local calendar'; *parapēgmata* are not calendars ('conventions for the dating of events'), but may incorporate calendars. Neugebauer (1975) 2: 587 uses the term 'calendaric text', but notes, 2: 589, that *parapēgmata* are 'also called "calendars," but not to be confused with Greek civil calendars'. Acknowledging these distinctions, I will occasionally use the terms 'calendar' and 'almanac' to refer to lists of astronomical events correlated with meteorological phenomena and other events (including, for example, agricultural tasks); I also recognize that the terms 'farmers' calendar' and 'agricultural almanac' are conventionally used to refer to many of the texts considered here. I thank Dr Lehoux for sharing his work with me.

2 Sider (2002b) 100. I am grateful to Professor Sider for sharing a pre-publication version of his paper.

3 Quoted in Swerdlow (1998) 2. See also Brown (2000).

4 Swerdlow (1998) 5, 17 and *passim*.
5 See Hunger (1977); Swerdlow (1998) 16–18. Brown (2000) 29, 97 and 99, and Swerdlow both suggest that frequent bad weather may actually have spurred the development of mathematical astronomy in Babylonia.
6 Graßhoff (1999).
7 See, e.g., Toomer (1988).
8 The omen literature suggests to some scholars the possibility that such a view was also held by Babylonians, but no 'theoretical' statement of such a view is known. See also Lloyd (1996a) chap. 8 ('Heavenly harmonies'), 165–89.
9 See ἁρμονία in LSJ. There are references to Pythagorean theory in Aristotle's *On the Heavens* 290b13 and in the text *On the Cosmos* 399a12, sometimes attributed to Aristotle. See also Huffman (1993).
10 Plato *Timaeus* 90D, trans. Jowett. The word 'harmonia' appears in the sentence which follows that quoted above: 'These each man should follow, and by learning the harmonies and revolutions of the universe, should correct the courses of the head which were corrupted at our birth. . . .' Important as Plato is as a source of knowledge about Pythagoreanism, I do not wish to give the impression that the *Timaeus* should be read only in that way. Plato is doing something which is distinctively his; see Sedley (1997).
11 This synopsis is based on that of Evelyn-White (1914) xvii–xviii. For an alternative breakdown of the poem into three sections, see Thalmann (1984); see also Nelson (1998).
12 Some scholars, including Fränkel (1975) 112 and Toohey (1996) 23, have suggested that the 'calendar' may originally have been a separate work not written by Hesiod; but West (1978) 41–59 argues that Hesiod composed it himself.
13 Trans. Lattimore (1991) 85, 89–91. See West (1978) 253 for an outline of Hesiod's 'calendar' and 380–1 for an explanation of his calculations.
14 Trans. Evelyn-White (*LCL*) 49 and 51; Lattimore (1991) 85 more literally translates μετὰ τροπὰς ἠελίοιο as 'after the sun has turned in his course'.
15 J. Pinsent (1989) 33–4 has suggested that the so-called 'Farmer's Year' contained in the final section of the *Works and Days* (lines 383–617) actually contains two different 'calendars': 'an astronomical one, based on the appearance of a limited number of stars and constellations, and a natural one based on the appearance and behaviour of birds, insects and flowers'. So, for example, 'the crane and cuckoo mark ploughing, the swallow the pruning of vines, and the snail the end of their cultivation and the beginning of harvest'. Pinsent argues that in both cases the calendars present a minimal amount of information; the calendars may represent traces of traditional Boeotian calendar poetry. See also Mair (1908), in an addendum to his translation entitled 'The Farmer's Year in Hesiod'; Stokes (1962, 1963).
16 Another Hesiodic work, the *Astronomy* (of which only fragments survive), may also have included a sort of almanac. Evelyn-White (1914) xix–xx suggests that the work 'gave some account of the principal constellations, their dates of rising and setting, and the legends connected with them, and

probably showed how these influenced human affairs or might be used as guides'. See also West (1978) 23. West (1996) noted that the *Works and Days* 'has closer parallels in near eastern literatures than in Greek, and seems to represent an old traditional type'. Bowen and Goldstein (1988) 57–8 emphasize the Near Eastern roots of the *parapēgma* tradition, which they trace to the omen and wisdom literature.

17 Diels and Rehm (1904). See also Rehm (1904); Dessau (1904); Evans (1983) A-580.

18 The so-called 'Geminus *parapēgma*' will be discussed below.

19 The noun is related to a verb meaning to 'fix beside, or near'; see also Neugebauer (1975) 2: 587 n. 3 and LSJ, s.v. παραπήγνυμι. A fragment of the earlier Milesian stone *parapēgma* is understood to have included directions for use; the word '*parapēgma*' is thought (by reconstruction) to have been part of the inscription. On the partially preserved directions for using the *parapēgma* and instructions for placing the peg, see Diels and Rehm (1904); Evans (1983) A-579–80 discusses the use of the infinitive in the directions for placing the peg. See also Dicks (1970) 84–5 on *parapēgmata*. LSJ suggest that the use of the verb in the pseudo-Platonic *Axiochus* (370c) indicates a familiarity with *parapēgmata*.

20 Neugebauer (1975) 2: 587. See also Evans (1983) A-578; Evans (1998) 201; Pritchett and van der Waerden (1961) 40 for mention of public use.

21 Several other fragments are known and are mentioned by Neugebauer (1975) 2: 587. See also A. Rehm (1949); Degrassi (1963) section on *parapēgmata* [299ff.]; Lehoux (2000).

The Milesian fragments are the best preserved examples of inscribed stone *parapēgmata* which detail astronomical phenomena. Rehm and, following him, Neugebauer include two small fragments, one from Athens (Kerameikos = A3 in Rehm), the other from Pozzuoli (Puteoli = A4 in Rehm), in their discussion of *parapēgmata*. It is not clear that the fragment from Kerameikos is astronomical; I thank Robert Hannah and Joyce Reynolds for sharing their views with me on this. The fragment from Puteoli does appear to be astrometeorological. On the Kerameikos fragment, see Brückner (1931) 23, and Kirchner (1931) 785, item 2782. On the Puteoli fragment, see Mingazzini (1928); Degrassi (1963) number 57 (Parapegma Puteolanum), 310–11.

22 Cicero *Letters to Atticus* 14 and 15; E.O. Winstedt (*LCL*) 373 in his translation of Letter 14, explains that the phrase παράπηγμα ἐνιαύσιον (*Letter* 14) corresponds to the phrase *clavus anni* in *Letter* 15, and that 'the expression arose from the custom of driving a nail into the right wall of the Temple of Jupiter on the Ides of September every year to keep count of the years'. G.R. Mair (1921) *Aratus* (*LCL*) 205 notes that 'it was usual for early astronomers to "fix up," παραπηγνύναι, their calendars on pillars in a public place; hence παράπηγμα, *affiche*, comes to mean "calendar"'. (Mair cites Aelian *Varia Historia* 10.7 which contains a reference to pillars (στῆλαι) set up by Meton, to record the solstices.) Socrates, in the pseudo-Platonic dialogue *Axiochus* 370c, suggests that the production of *parapēgmata* is one of several special human activities linked to the divine spirit (πνεῦμα θεῖον).

23 One *parapēgma* fragment contains partially preserved instructions for use. Diels and Rehm (1904) 102–3 link the fragment (456C) to the earlier *parapēgma*; Rehm (1949) cols 1299–301, revising his earlier view, suggests that the fragment is related to the later *parapēgma*. Cf. Evans (1983) A-580; Lehoux (2000) 32.

24 On ancient calendars, see Bickerman (1980) 13–61.

25 Aristophanes, in the *Clouds* (614ff.), joked about the Athenian calendar and attributed to the moon complaints that the Athenians were not arranging the days properly in accordance with the lunar phases. See also Meritt (1928).

26 Neugebauer (1975) 2: 588. Neugebauer argues (2: 617, 622) that 'the primary purpose of the known [calendaric] cycles from the 5th and 4th centuries was the construction of reliable parapegmata, not the reform of civil calendars'. In the view of Toomer (1996c), Meton and Euctemon were not interested in calendar reform, but rather in providing a fixed basis for astronomical observations. For a detailed discussion of Meton as a parapegmatist, see Bowen and Goldstein (1988). Hannah (1999) has argued that Meton and Euctemon's purpose may well have been to do with calendar reform.

27 I thank Robert Hannah for his help here. Regarding the size of the *parapēgmata*, as an indication, surviving fragment 456B, described below and illustrated, is 0.44 metre wide and about 0.26 metre high; 456A (also described and illustrated) is 0.54 metre wide, 0.22 metre high and about 0.18 metre thick.

28 Diels and Rehm (1904) 104; cf. Evans (1998) 202. The term 'acronychal' is often defined as 'at nightfall'; Evans (1998) 197 and North (1994) 52 point to the ambiguity of the term, noting that it may refer to either the first or last evening sighting (cf. *Compact Oxford English Dictionary*).

29 Neugebauer (1975) 2: 588 described the earlier *parapēgma* as 'peculiar also in so far as it ignores the meteorological purpose of such a calendar', apparently here not considering winds to be meteorological. See Virgil *Georgics* 1.50ff. and also Pliny *NH* 18.56.205–6 on the importance of knowledge of the winds. Cf. also the entries for the Water-Pourer (Aquarius), days 14–17, in the 'Geminus *parapēgma*', in Geminus *Elementa astronomiae* (ed. Manitius) 226–7, which mention the winds. The 'Geminus *parapēgma*' will be discussed more fully below.

30 There have, for example, been attempts to reconstruct Euctemon's *parapēgma*, based on fragments preserved on later texts; see, for example, Rehm (1913) and Hannah (2002). I thank Professor Hannah for sharing his paper with me.

31 Diels and Rehm (1904) 110. Less of this inscription has survived; for legibility here, indications of the inferences made by Diels and Rehm have not been included.

32 The literary form simply gives the daily progress of the sun, since no holes would have been cut in the papyrus roll; cf. the 'Geminus *parapēgma*' described below. It should be noted that these texts contain no explanation of the physics that mediates between celestial and terrestrial events; such explanations must be sought in other genres.

33 Joyce Reynolds (personal communication) has quite rightly suggested that the location of the finds of the fragments should be more closely investigated. Harry Pleket (personal communication) suggested that the *parapēgmata* would have probably had some official status.

34 Cf. Lehoux (2000) 123–4 and n. 8.

35 See Mingazzini (1928); Rehm (1949); Degrassi (1963); Lehoux (2000) 33–4.

36 It is difficult to give an exact figure for the number of *parapēgmata* which survive, because there is some latitude in defining what should count as a *parapēgma* and what constitutes survival, when one speaks of ancient texts. See, again, Lehoux (2000) *passim.*

37 The district of Sais and its great city, known by the same name, are described by Plato in the *Timaeus* 21e–22a.

38 Fowler (1999) 229–30. See also Fowler and Turner (1983). The dating of the text is based on internal evidence, and also palaeographical analysis.

39 P. Hibeh 27, in Grenfell and Hunt (1906) 151. Grenfell and Hunt provide a transcription and translation of the text; see also Fowler (1999) 229–30 on Grenfell and Hunt; Neugebauer (1975) 2: 599–600 and 2: 688–9; Lehoux (2000) 31.

40 See Evans (1983) A-569, also Evans (1998) 199, and Aujac (1975) 157.

41 On the dating of Geminus, see Jones (1999). On the dating of the *parapēgma,* see Neugebauer (1975) 2: 580–1; Evans (1983) A 576–7; see also Evans (1998) 199, and Toomer (1996b). Aujac (1975) 157–8 holds that the *parapēgma* was the work of Geminus; cf. Lehoux (2000) 34. For the text of the 'Geminus *parapēgma*', see Geminus *Elementa astronomiae* (ed. Manitius) 210–33, and *Géminos* (ed. Aujac) 98–108.

42 Χρόνοι τῶν ζῳδίων, ἐν οἷς ἕκαστον αὐτῶν ὁ ἥλιος διαπορεύεται, καὶ αἱ καθ' ἕκαστον ζῴδιον γινόμεναι ἐπισημασίαι, αἳ ὑπογεγραμμέναι εἰσίν. The translation here follows that of Aujac (1975) 98; cf. Neugebauer (1975) 2: 588: 'Time intervals in which the sun traverses the single signs with prognostications as recorded.' (As Neugebauer (1975) noted (2: 588 n. 6), 'For reasons unknown Manitius . . . translated neither the title nor the headings but gave only a paraphrase.')

43 'Geminus *parapēgma*', text in Manitius (1898) 224–6; Aujac (1975) 104–5.

44 See Rehm (1949) 1305–6; Hannah (2002); Evans (1998) 199; van der Waerden (1988) 77.

45 I thank Geoffrey Lloyd for pointing out that there are two levels of compilation at work in the Hippocratic corpus. Some treatises, e.g. *On the Nature of Man*, are themselves compilations (perhaps by different authors, but certainly of disparate materials). Second, the compilation of the corpus as a whole involved the collection of the various individual works.

46 Also mentioned by the author of *On Weather Signs* 1.4.

47 Aujac (1975) 157. On Democritus, see Sider (2002a). Diogenes Laertius 9.48 gives the title: Μέγας ἐνιαυτὸς ἢ 'Αστρονομίη, παράπηγμα for Democritus' *parapēgma.*

48 Robert Hannah (personal communication) has raised the question as to whether the 'predictions' were a relic from a (Near Eastern) ancestor, in which they served some 'ominous' purpose. The relationship between the Greek practice of correlating stellar phases with weather phenomena, in the *parapēgmata*, and the Babylonian astronomical diaries is still to be investigated.

49 Wood (1894) 11 argued that 'On Weather Signs' 'reads as if it were a collection of notes taken of a lecture delivered, or notes to form the basis of, and to be expanded in, a lecture to be delivered'. He suggested that the work may 'be a *resumé* of what he [Theophrastus] heard from his master [Aristotle]'.

50 Cronin (1992) 335–7.

51 Sider (2002b) 108.

52 A work by Aristotle *On Signs of Storms* (σημεῖα χειμώνων) is listed by Diogenes Laertius 5.26; see Sharples (1998a) 145 n. 389. Diogenes Laertius 5.45 mentions Theophrastus' περὶ σημείων; cf. 137.17 in FHS&G 1: 282.

53 *On Signs* 11, trans. Sider (forthcoming). I am grateful to Professor Sider for sharing a pre-publication version.

54 Sider (2002b) 107.

55 Cronin (1992) 317, also 313, 335.

56 Cronin (1992) 323, 326ff. Cronin (1992) 326 notes that only two signs given in the work fall outside of these categories, namely, bubbles rising on rivers (section 13) and the motion of cobwebs (29).

57 *On Signs* 1, trans. Sider (forthcoming). A number of alternative readings have been offered for the phrase ἐκ τῶν ἀστρονομικῶν; Hort (1916–26, *LCL*) 2:391 suggests 'from astronomy' or 'from my astronomical works', while Wood (1894) 53 offers 'the information of astronomers'. Cronin (1992) 313 suggests that the phrase should be read specifically as referring to 'either a specific parapegma or to parapegmata in general for information which he himself will not provide'. He adds that 'other possible but less likely meanings' of the phrase are 'from works on astronomy' or 'from astronomy'. It is not entirely clear that the phrase need refer to *parapēgmata*; that the word *parapēgma* is not used may indicate that more general astronomical works were meant. Sider, in his (forthcoming) commentary on *On Signs*, argues that 'the sense is hardly affected, however the word is taken'.

58 Sider (2002b) 102 and n. 16. See also Lloyd (1996b) 162f.

59 Trans. Wood (1894) 54. See also Tannery (1893) 17–19 and Bowen and Goldstein (1988) 80 on what sort of observations may have been made by these astronomers. As Bowen and Goldstein note, there is some ambiguity regarding the solstitial observations, whether there were single observations made by each astronomer, or whether they engaged in a programme of observation. Robert Hannah has noted (personal communication) that Mt Lykabettos is really just a hill; Paul Cartledge mentioned its height and steepness.

60 See Sider (2002b) 104.

61 Cf. Rehm (1941) 137. Cronin (1992) 322 notes that 'this section has but the most tenuous connection with the main theme'. He suggests that

'its purpose is probably to show how different localities, just as they may produce experts in astronomy, because of the presence in them of mountains, so too may produce experts in weather prediction'.

62 *On Weather Signs* 1.1 and 1.4.

63 Goldstein and Bowen (1983) 340. On 331 they suggest that 'Greek astronomy, as the very name implies, began as the organization of the fixed stars into constellations'. In their view, 'the purpose was to construct a calendar by correlating dates and weather phenomena with the risings and settings of the fixed stars or constellations'; 'the astronomical calendar of risings and settings, weather phenomena, and seasons took the form of a *parapēgma*'.

64 Evans (1983) A-582–4 notes that Ptolemy innovated by focusing on fifteen stars of the first magnitude and fifteen of the second, rather than on constellations. Ptolemy also specified the latitude for each star phase on a given date but does not seem to have added any weather predictions.

65 On the meaning of *klima* see Toomer (1984) 19; Evans (1998) 94.

66 Cf. Neugebauer (1975) 2: 929, and Lehoux (2000) 130–45 on the meaning of ἐπισημαίνει.

67 Ptolemy *Phases* (ed. Heiberg) 26–7, trans. based on Evans (1998) 203–4, with modifications; see also Neugebauer (1975) 2: 928–9.

68 Evans (1983) A-582–3. On the 'Egyptians', see Neugebauer (1975) 2: 562–3; van der Waerden (1985).

69 On Dositheus, see Netz (1998).

70 See Evans (1983) A-582–4. On Sosigenes, see Neugebauer (1975) 2: 575.

71 This is, for example, a characteristic of Aristotle's approach, although, as David Sedley reminded me, it should be acknowledged that Aristotle often cites predecessors in a relatively friendly way, when gathering *endoxa*.

72 I thank Harry Pleket for this suggestion.

73 I thank Marlein van Raalte for bringing this to my attention, in this context.

74 Thomas (1992) 84. The question of literacy, as it relates to inscriptions generally and to *parapēgmata* specifically, is somewhat problematic; see Harris (1989).

75 Judge (1997) 807–8.

76 Pliny *NH* 18.57.210–17. Pliny (18.60.224–18.74.320) presents an 'almanac', based on a number of sources. Agricultural advice is included, along with the star phases and weather predictions. His contemporary, Columella (*On Agriculture* 11.2.3–98), provided the traditional phases and weather predictions, followed by appropriate agricultural advice. Columella does not list authorities here, which is not surprising given his view that weather cannot be predicted according to fixed dates (11.1.31–2.2). He did, however, cite authorities at the beginning of 1.1. Varro, an earlier agricultural author, provided a brief description of the seasons, with fewer astronomical and meteorological phenomena detailed (*On Agriculture* 1.28–37); cf. Lehoux (2000) 76.

77 See, for example, Pliny *NH* 18.69.290. Furthermore, in his brief statement at the opening of *On Weather Signs* (1.1), the author indicated that 'those signs which belong to the setting or rising of the heavenly bodies must be learnt ἐκ τῶν ἀστρονομικῶν'; whether the latter phrase refers to astronom-

ical works or to *parapēgmata*, these' works would be the productions of specialists. Cronin (1992) 313 suggests that the author of *On Weather Signs* may have wished to distinguish his own work from these others.

78 Thomas (1992) 85.

79 Cf., e.g., the weather phenomena in the Homeric poems, which often appear as the epiphanies of gods, who are capricious and arbitrary and anything but rational and predictable.

80 Pliny *NH* 18.57.210–17. Cf. also *On Weather Signs* 1.3 with regard to the importance of considering the specific location of observers of weather signs.

81 Ptolemy *Syntaxis Mathematica* 7.3 (ed. Heiberg 2: 17–18), trans. Toomer (1984) 329.

82 Ptolemy *Phases* (ed. Heiberg) 66–7; Evans (1983) A-582–3.

83 Ptolemy's sensitivity to the importance of location is indicated, as well, in his work on geography and in his view (*Tetrabiblos* 2.2) that living in particular places, and experiencing particular atmospheric effects, will have particular effects on the population.

84 Neugebauer (1975) 2: 589 n. 19 and 926 n. 4.

85 On Ptolemy's views on the stochastic character of some types of astronomical work, see Taub (1997) 86.

86 *Syntaxis* 8.6 (ed. Heiberg 2: 204), trans. Toomer (1984) 417.

87 Neugebauer (1975) 1: 141; see also Ptolemy *Tetrabiblos* 2.12 and 13, and also Toomer (1984) 417 n. 214.

88 Ptolemy *Syntaxis* 6.7 (ed. Heiberg 1: 512).

89 Ptolemy *Syntaxis* 6.11 (ed. Heiberg 1: 536–7).

90 Ptolemy's desire to provide others with easy means of calculation is also demonstrated by his provision of the so-called *Handy Tables*; see Pedersen (1974) 397ff. for a discussion of this work.

91 Graßhoff (1993) has analysed the relationship between Ptolemy's *Phases* and his theoretical considerations in the *Syntaxis*; he has concluded that the *Phases* was written later. Graßhoff argues that Ptolemy uses intrinsic Babylonian values, showing that he considered Babylonian observations and theories as fundamental for the derivation of stellar phases. The modes of transmission and acquisition of Babylonian astronomical and meteorological knowledge are still being studied, but the reliance of Greek astronomers on Babylonian material is widely acknowledged. See, e.g., Toomer (1988) and Jones (1991).

92 Ptolemy elaborated (*Tetrabiblos* 1.1, trans. Robbins (*LCL*) 3):

> Of the means of prediction through astronomy, O Syrus, two are the most important and valid. One, which is the first both in order and in effectiveness, is that whereby we apprehend the aspects of the movements of sun, moon, and stars in relation to each other and to the earth, as they occur from time to time; the second is that in which by means of the natural character of these aspects themselves we investigate the changes which they bring about in that which they surround.

See also Long (1982) esp. 178–83.

93 Ptolemy *Tetrabiblos* 1.1, trans. Robbins (*LCL*) 5.

94 Ptolemy *Tetrabiblos* 1.2, trans. Robbins (*LCL*) 9–11. It is significant that observation alone does not qualify as natural philosophy.

95 Ptolemy *Tetrabiblos* 1.2, trans. Robbins (*LCL*) 11.

96 Ptolemy *Tetrabiblos* 1.2, trans. Robbins (*LCL*) 11–13.

97 Ptolemy *Tetrabiblos* 2.13, trans. Robbins (*LCL*) 213, with modifications. Lehoux (2000) 131ff. argues that the plural noun *episēmasiai* should be understood, in the astrometeorological literature, as 'changes in the weather'; cf. LSJ, s.v. 'ἐπισημασία', III.2. The text here (unlike the *parapēgmata*) assumes that observations will actually be made.

98 Ptolemy *Tetrabiblos* 2.13, trans. Robbins (*LCL*) 213–15.

99 Ptolemy *Tetrabiblos* 2.13, trans. Robbins (*LCL*) 216–17.

100 Ptolemy detailed the role of individual planets with regard to winds at 1.18.

101 Ptolemy *Tetrabiblos* 2.13; cf. *On Weather Signs* 13.

102 Ptolemy *Tetrabiblos* 2.13, trans. Robbins (*LCL*) 199. Robbins suggests that 'by [the phrase] "the more general natures" doubtless are meant temperature and other things, besides the winds, that go to make up the weather'.

103 Ptolemy *Tetrabiblos* 1.1.

104 Ptolemy presents his ideas on cosmic harmony in the *Harmonics* 3.8–16. See Ptolemy *Harmonics* 3.12, in which the modulations of *tonoi* correspond to lateral movements of the stars. But note that the state of the text is problematic; see Lloyd (1996a) chap. 8, particularly 178–9. See also Barker (2000) 267–9.

105 Pliny *NH* 2.39.105–2.41.110, trans. Rackham (*LCL*) 249.

106 Geminus *Introduction to the Phenomena* (17.26ff.); Lewis (1862) 311.

107 Epicurus 'Letter to Pythocles' in Diogenes Laertius 10.98–9, trans. Hicks (*LCL*) 2: 627.

108 Cf. Sextus Empiricus and Cicero below; also cf. Pliny, e.g. *NH* 18.57.211.

109 Columella *On Agriculture* 11.1.30–2, trans. Forster and Heffner (*LCL*) 3: 69. Cf. Geminus *Introduction* 17.23ff.

110 Columella *On Agriculture* 11.2.2, trans. Forster and Heffner (*LCL*) 3: 71.

111 The seasonal information given by Marcus Terentius Varro in *On Agriculture* 1.28–37 is linked to astronomical events. Cato also uses astronomical time-keeping, e.g. *On Agriculture* 44: 'The trimming of the olive-yard should begin fifteen days before the vernal equinox; you can trim to advantage from this time for forty-five days', trans. Hooper and Ash (*LCL*) 63.

112 Pliny *NH* 18.60.225, trans. Rackham (*LCL*) 331–3.

113 Sextus Empiricus *Against the Professors* 5.1–2, trans. Bury (*LCL*) 323. Cf. Lewis (1862) 311.

114 Geminus *Introduction* 17.23ff. (ed. Manitius 188–90); see also Lewis (1862) 311. Cf. Ptolemy *Tetrabiblos* 1.1, cited above.

115 Geminus *Introduction* 17.26–30 (ed. Manitius).

116 See Lehoux (2000) 34–5.

117 See Neugebauer (1975) 2: 581 and 583 on Geminus and his views against astrology and *parapēgmata*.

118 Harris (1989) 11, 111–12 and 53–5. I thank Robert Hannah and Paul Cartledge (personal communication) for the point about 'mass-production' of *ostraka*.

119 Harris (1989) 130, citing *Sylloge Inscriptionum Graecarum* (3rd edn) 577; Harris also mentions the edition and commentary by E. Ziebarth, *Aus dem griechischen Schulwesen* (2nd edn, Leipzig and Berlin 1914) 2–29.

120 Harris (1989) 133.

121 Harris (1989) 201 n. 132, citing L. Bove, *Documenti di operazioni finanziarie dall'archivio dei Sulpici* (Naples, 1984) 100–5. There is also evidence that third parties would help those who could not read with their financial matters; Harris 262 n. 454, citing G. Pupura in *Atti del XVII Congresso internazionale di papirologia* (Naples, 1984) iii.1248–54, offers an example of an illiterate who was involved in financial transactions in Puteoli in the year 38.

122 Harris (1989) 239 n. 334.

123 Hannah (2001) 155–9.

124 Pliny the Elder also produced an agricultural almanac in his *Natural History* (discussed in the final chapter here). Beagon (1989) argues that Pliny's equestrian and agricultural commitments should be taken seriously. Lehoux (2000) 36 counts Pliny's almanac as a *parapēgma*.

125 Hannah (2001) 151 has argued, contra Goody (1977), that the compiling of lists (including those, like *parapēgmata*, which incorporate variable numerical values) is not an activity restricted to the literate. See also Sider (2002b) 109.

126 Sider (2002b) 104.

127 Trans. Sider (forthcoming).

128 As was noted earlier, Cronin (1992) has produced a detailed analysis of the categories of signs present in the work. The sign of bubbles on the river is one of only two (the other being the motion of cobwebs) which do not fit into any of Cronin's categories.

129 Here, I rely heavily on the references provided by Hort (1916–26), Mair (1921), Fairclough (1916) and Rackham (1950, Vol. 5). For frogs as 'weather-prophets' (Falconer's (1923) 238 n. 3 phrase), see Aratus 948; Pliny 18.361–2; Virgil *Georgics* 1.378. See also the commentaries of Kidd (1997) and Sider (forthcoming).

130 Diogenes Laertius 5.26 and 5.45.

131 Schiesaro (1996). Harris (1989) 49 endorses the view of West (1978) 25–30 that the *Works and Days* 'is to be seen not as a didactic poem for farmers but rather as an example of wisdom literature'. However, the term 'didactic poem' may be applied to works of 'exhortation and instruction' (West's phrase); the poem may be regarded as didactic, without suggesting that it was intended to be read by farmers.

132 Dalzell (1996) 26. As he explains, while Hesiod may well have had a brother named Perses, 'the contradictions and inconsistencies in the story suggest that the Perses of the poem is at least in part a figure of convention'. See also West (1978) 33–40; Lamberton (1988) 23–4; Clay (1993).

133 See Bing (1993) esp. 103.

134 Aratus *Phaenomena* 1–18, trans. Poste (1880) 1–2. I have chosen this translation to give a sense of the poetic form of the work; further translations of *Phaenomena* used here are by Kidd (1997).

135 Kidd (1997) 162.

136 Toomer (1996a). See D. Kidd (1997) 425 on the title of the section on signs.

137 Toohey (1996) 1; see Maass (1898).

138 Toomer (1996a); Ptolemy *Syntaxis*, e.g. 3.1, (ed. Heiberg 1: 191, 200).

139 Mair (1921) (*LCL*) 204. Mair 204–5 discusses the 'vexed question' of the relation of this section of the poem to that of the work known as *On Weather Signs*.

140 Toohey (1996) 53.

141 Trans. Kidd (1997) 127.

142 Toohey (1996) 57, who follows Effe (1977) 40–56. Cf. James (1972).

143 Toohey (1996) 61.

144 Aratus *Phaenomena*, trans. Kidd (1997) 129. The 'nineteen cycles' in the first part of this passage is very likely a reference to the nineteen-year cycle that Meton developed to deal with the problem of co-ordinating solar years and lunar months; he found that 235 lunations practically coincide with nineteen solar years. See Toomer (1974). The following lines seem to refer not to the Metonic cycle but rather to the year. See also Mair (1921) 205.

145 See also Sider (2002a) 103.

146 Aratus *Phaenomena*, trans. Kidd (1997) 131.

147 Mair (1921) *Aratus* (*LCL*) 266–7 points to these corresponding passages; it is worth comparing Pliny *NH* 18.58.218 until the end of the book, as well.

148 Aratus *Phaenomena* 909–12, trans. Kidd (1997) 139.

149 Aratus *Phaenomena* 913–19, trans. Kidd (1997) 139–41.

150 Aratus *Phaenomena* 1064–7, trans. Kidd (1997) 151. Compare the similar passage in *On Weather Signs* 47.

151 Trans. Kidd (1997) 141.

152 Several of the editors of these authors have made extensive notes regarding the parallels and similarities among the descriptions of weather signs; the discussion here is meant simply to give an indication, rather than an exhaustive listing.

153 Dalzell (1996) 26–7.

154 Dalzell (1996) 27.

155 Dalzell (1996) 50. See also Bing (1993) esp. 103.

156 See Courtney (1996); Blänsdorf (1995) 238–9; Courtney (1993) 244–5.

157 Germanicus' authorship was accepted by Le Boeuffle (1973) but queried by Gain (1976) 16–20, who argues that the evidence suggests that the poem may have been composed either by Germanicus or by his uncle Tiberius.

158 Gain (1976) 13. Courtney (1996) notes that the 200 lines in Germanicus on the sidereal influences on weather (not found in Aratus) may have been intended for a separate work. See also Housman (1900).

159 Germanicus *Aratus* frag. iii lines 1–9, trans. Gain (1976) 73.

160 Germanicus *Aratus* 5ff., trans. Gain (1976) 53.

161 Toohey (1996) 187–8 citing relevant passages in Germanicus' poem.

162 Whether or not this Avienus was the author of the *Fables* is not entirely clear; see Duff and Duff (1935) 669. See Cameron (1967, 1995) on the question of Avienus' identity.

163 See Soubiran (1981).
164 Zehnacker (1989) 328. See Müller (1861) vol. 2 for Dionysius Periegetes' geographical poem; van de Woestijne (1961) has edited Avienus' *Descriptio Orbis Terrae.*
165 Soubiran (1981) 41–2; Zehnacker (1989) 325.
166 Soubiran (1981) 61ff.
167 Fowler and Fowler (1996b). See also Hardie (1986) 157–67 and, more generally, Farrell (1991).
168 Nelson (1998).
169 Jermyn (1951). See also Wilkinson (1997) 234–42.
170 *Georgics* 1.316ff., trans. Day Lewis (1940/1999 reprint) 61–2; 1.334 and 1.351ff., trans. Fairclough (*LCL*) 105.
171 Trans. Day Lewis (1940/1999) 63.
172 Trans. Day Lewis (1940/1999) 63.
173 *Georgics* 1.373ff., trans. Day Lewis (1940/1999) 63–4; 1.390 trans. Fairclough (*LCL*) 107.
174 Trans. Day Lewis (1940/1999) 67.
175 Fairclough (1916) 113 n. 1.
176 Dalzell (1996) 110–11.
177 Toohey (1996) 23.
178 Schiesaro (1996). See also Schiesaro *et al.* (1993).
179 I thank David Sedley for discussing this point with me.
180 Other passages from Cicero's translation of Aratus may be found in his *On the Nature of the Gods* (= *De natura deorum*). See also Mansfeld (1992b) 327–8; Allen (2001) 162ff.
181 Cicero *On Divination,* trans. Falconer (*LCL*) 223.
182 Trans. Falconer (*LCL*) 227.
183 Trans. Falconer (*LCL*) 231.
184 Trans. Falconer (*LCL*) 237.
185 Presumably this is a reference to Cicero's translation of the second part of Aratus' poem, and not an entirely separate work. Trans. Falconer (*LCL*) 237.
186 Cicero *On Divination* 1.7–9, trans. Falconer (*LCL*) 237–9.
187 Cicero *On Divination* 1.58, trans. Falconer (*LCL*) 369.
188 Cicero *On Divination* 2.5, trans. Falconer (*LCL*) 383.
189 Cicero *On Divination* 2.5, trans. Falconer (*LCL*) 385.
190 Cicero *On Divination* 2.6, trans. Falconer (*LCL*) 387.
191 Vitruvius *On Architecture* 9.6.3, trans. Granger (*LCL*) 2: 247. Aratus is included in the list of parapegmatists by Vitruvius (9.6.3). On this, see Gee (2000) 19–20 and Soubiran (1969) referring to 9.6.3.
192 See Denyer (1985) who discusses Cicero, Stoicism and the differing predictive methods of ancient science and divination; and, contra Denyer, Timpanaro (1994). See also Long (1982) 165–92; Barton (1994a, 1994b); Beard (1986); Schofield (1986).
193 Cicero *On Divination* 1.1, trans. Falconer (*LCL*) 223–5.
194 Cicero *On Divination* 2.42, trans. Falconer (*LCL*) 471–3.
195 Cicero *On Divination* 2.43, trans. Falconer (*LCL*) 473.
196 Cicero *On Divination* 2.45, trans. Falconer (*LCL*) 477.
197 Pliny *NH* 2.6.29, trans. Rackham (*LCL*) 189.

198 Pliny *NH* 2.41.108–10, trans. Rackham (*LCL*) 251–3.
199 Pliny's discussion of celestial influence on weather will be addressed again in the final chapter.
200 French (1994) 17.
201 Scholfield (1958) (*LCL*) 2: 103 notes that this reference must be to *On Weather Signs*.
202 Aelian *On the Characteristics of Animals* 7.7–8, trans. Scholfield (*LCL*) 2: 103, 105–7.
203 Trans. Scholfield (*LCL*) 2: 107–9. On Anaxagoras' rain-wear, see also Diogenes Laertius 2.10.
204 Linderski (1996).
205 Cicero *On the Nature of the Gods* 2.5, trans. Rackham (*LCL*) 137.
206 French (1994) 165f.
207 Aristotle *Problems* 26.23 (942b16), trans. Forster (*Complete Works*) 2: 1478.
208 Another indication of the level of ancient familiarity with weather signs is evident in Matthew's report of Jesus' response to the request from the Pharisees and the Sadducees for a sign from heaven:

> He answered and said unto them, When it is evening, ye say, *It will be* fair weather: for the sky is red.
> And in the morning, *It will be* foul weather to day: for the sky is red and lowring. *O ye* hypocrites, ye can *discern* the face of the sky; but can ye not discern the signs of the times?
> (Matthew 16.2–3, King James Version of the New Testament.)

I thank Clare Drury, Diana Lipton and Talel Debs for discussing biblical weather signs with me.
209 See Bandy (1983) xxvii–xxviii.
210 See Maas (1992) 84.
211 See Maas (1992) 56 for a synopsis.
212 The summary provided here is based on that of Bandy (1983) xxix; cf. Maas (1992) 107.
213 Bandy (1983) xxix indicates that Lydus' work provides a historical survey of ancient astrology; Maas (1992) 105–7 describes Lydus as being particularly interested in astrology.
214 Maas (1992) 107.
215 Lydus *De Ostentis* 4, trans. Maas (1992) 106.
216 Lydus *De Ostentis* 9, trans. Maas (1992) 105.
217 Lydus *De Ostentis* 4, trans. Maas (1992) 105–6.

3 EXPLAINING DIFFICULT PHENOMENA

1 Pliny's *Natural History* is an important exception, and will be considered in the final chapter.
2 On the doxographical tradition, see KRS (1983) 4–6; see Mansfeld (1999a) 17–19 for a reappraisal of the use of the term 'doxography', and also Mansfeld and Runia (1997) xiii–xiv.

3 See, e.g., the discussion of the views of Anaxagoras, Democritus and Anaximenes on earthquakes, Aristotle *Meteorology* 2.7 and 2.8. There is evidence that Democritus was interested in weather prediction; see Sider (2002a).

4 Seneca *Natural Questions* (= *NQ*) III.14, trans. KRS (1983) 93. G.S. Kirk, KRS (1983) 93 suggests that this passage is 'presumably from Theophrastus, through a Posidonian source'.

5 Seneca *NQ* III.14, (*LCL*) 1: 230.

6 Lloyd (1970) 9.

7 *Suda* s.v. ᾿Αναξίμανδρος Πραξιάδου Μιλήσιος = KRS 95 (1983) 100.

8 Kirk, in KRS (1983) 138.

9 Hermann Diels conjecturally reconstructed Aëtius' text from later material transmitted under the names of Plutarch and Stobaeus, among others; see Diels (1879). On Aëtius, see the work of Jaap Mansfeld and David Runia, who have emphasized the diffusion and fluidity of the doxographical tradition, Mansfeld and Runia (1997) 330; Runia (1989); Mansfeld (1990).

10 Aëtius 3, 3, 1–2 = KRS 130, trans. KRS (1983) 137–8, slightly modified.

11 Seneca *NQ* II.18 = KRS 131, trans. KRS (1983) 138, slightly modified.

12 Diogenes Laertius 2.3 = KRS 138 (1983) 143.

13 Aristotle *Metaphysics* A3 984a5.

14 Kirk, in KRS (1983) 146, notes that 'ἀήρ in Homer and sometimes in later Ionic prose meant "mist", something visible and obscuring; and Anaximander's cosmogony included a damp mist, part of which congealed to form a slimy kind of earth'. See also KRS (1983) 132, 142.

15 Simplicius' commentary on Aristotle's *Physics* (*in Aristotelis Physica commentaria* = *in Phys.*) 24, 26 = KRS 140, trans. KRS (1983) 145. Hippolytus *Refutation of all Heresies* 1, 7, 1 = KRS 141 (1983) 144–5 also discussed Anaximenes' ideas regarding air and the various meteorological phenomena associated with its changes:

> It is always in motion; for things that change do not change unless there be movement. Through becoming denser or finer it has different appearances; for when it is dissolved into what is finer it becomes fire, while winds, again, are air that is becoming condensed, and cloud is produced from air by felting. When it is condensed still more, water is produced; with a further degree of condensation earth is produced, and when condensed as far as possible, stones.
>
> (Trans. KRS (1983) 145)

16 Aëtius 3, 3, 2 = KRS 159, trans. KRS (1983) 158.

17 In Aristotle's account of Anaximenes' views on earthquakes, it is not clear that the air plays a role: 'Anaximenes says that the earth, through being drenched and dried off, breaks asunder, and is shaken by the peaks that are thus broken off and fall in. Therefore earthquakes happen in periods both of drought and again of excessive rains; for in droughts, as has been said, it dries up and cracks, and being made over-moist by the waters it crumbles apart.' Aristotle *Meteorology* 365b6–12 = KRS 159, trans. KRS

(1983) 157–8. As Kirk points out, it is notable that air seems to play no part in the occurrence of earthquakes in Aristotle's account of Anaximenes' views, but this may be due to Aristotle's desire to distance himself from his predecessors.

18 Aëtius 1, 3, 4 = KRS 160, trans. KRS (1983) 158–9.

19 I have not attempted to provide here a survey of pre-Socratic meteorology. On Xenophanes' meteorology, see Mourelatos (1989).

20 Aristotle *History of Animals* 511b31 (DK64B6) = KRS 615, p. 450. See also Laks (1983).

21 Simplicius *in Phys.* 151, 20 = KRS 597. Kirk, in KRS (1983) 435, notes that *Meteorologia* is 'a highly dubious form of book-title'. He also notes, 436, that Diogenes may have been a medical practitioner; this may also have spurred his interest in meteorology. While the connection between medicine and meteorology is a topic outside the scope of this work, nevertheless it is worth mentioning several Hippocratic medical treatises in this context. The case notes provided in the *Epidemics* often incorporate comments on the weather conditions; see, e.g., 1.1–13. *Airs, Waters, Places* (1–2) emphasizes the influence of climate, water supply and location on human health and disease. The author of the *Humours* claims that it is not only possible to predict diseases from the weather (13ff.), but also possible to predict weather from diseases (17).

22 Theophrastus *Physical Opinions* (or *Opinions of the Physicists*) fr. 2 in Simplicius *in Phys.* 25, 1 (DK64A5) = KRS 598, KRS (1983) 436–7.

23 Aristophanes *Clouds* 227ff., trans. Sommerstein (1973) 121–2.

24 I am here primarily concerned only with the first three books of the *Meteorology*; I do not look in any detail at Book 4, which largely deals with the materials found in and around the earth. I have been guided by the second-century CE commentator Alexander of Aphrodisias, who explains that 'the book entitled "the fourth" of Aristotle's *Meteorology* does belong to Aristotle, but not to the treatise on meteorology, for the matters discussed in it are not proper to meteorology'. Alexander of Aphrodisias *On Aristotle Meteorology 4*, trans. Lewis (1996) 65.

Though the authenticity of Book 4 is disputed, I follow the view of Freeland (1990) 67 n. 1 that in choosing not to treat Book 4 now, 'this is due less to concerns about the authenticity of bk. 4 than to its fairly obvious lack of continuity with the earlier books in terms of approach and methods. The absence in bk. 4 of the fundamental *archai*, the exhalations, consistently used throughout 1–3, makes bk. 4 seem part of a different project (although it covers subject-matter we would expect to be included on the basis of the treatise's overall programme).' See also Sharples (1998a) 169–70.

For readers interested in Book 4, see the extensive bibliography in Barnes (1995a) 334–5, and also Lewis (1996) Introduction and Notes.

Sadly, due to limitations of space, the extant ancient commentaries on Aristotle's *Meteorology* (by Alexander, Olympiodorus and John Philoponus) must remain outside of the scope of the present work.

25 Cf. Capelle (1912). The description of Aristotle's works here is not meant to make any claim regarding the order of composition.

26 Aristotle *Meteorology* 338a20ff, trans. Lee (*LCL*) 5. I am generally, here, using the Lee translation (*LCL*), sometimes with modifications.

27 See Lee (1952) 36–7 on whether these phenomena are the aurora.

28 The use of the term *prēstēr* (πρηστήρ) is not straightforward. It is defined in LSJ as 'hurricane or waterspout attended with lightning' on the basis of its usage in several of the texts considered here: Aristotle *Meteorology* 371a16, Epicurus 'Letter to Pythocles' (Diogenes Laertius 10.105), Lucretius 6.424, 445. As Furley (1955) 368–9 pointed out, the author of *On the Cosmos* seems to use it to refer to a sort of whirlwind and also a kind of thunderbolt, while for Aristotle, in the *Meteorology* (371a15), the term seems to be associated with fire (as a fiery whirlwind). Sedley (1998a) 158–9 and 182 defines *prēstēr* as waterspout (treated by Lucretius after thunderbolts and before cloud), and he considers the order in which it is discussed as indicating Theophrastus' influence. Traditionally, the *prēstēr* was considered to be fiery (coming from above), but Theophrastus included it under the heading of winds (*Meteorology/Metarsiology* [13] 43–54, in Daiber (1992)), and in *On Winds* 53, Theophrastus seems to use the term to describe a waterspout; I thank Marlein van Raalte for her help here.

29 Aristotle *Meteorology* 338b25–339a2, trans. Lee (*LCL*) 5. See Hine (2002) 58–60 and 69 on differences in the treatment of earthquakes and volcanoes in the ancient literary traditions; generally, earthquakes were included within the meteorological literature, while volcanoes were not.

30 Aristotle *Meteorology* 344a5–7, trans. Lee (*LCL*), with modification.

31 Aristotle *Posterior Analytics* 1.2. See definition in Smith (1995) 47.

32 See Burnyeat (1981) for an account of the sort of explanation which provides 'understanding'. Aristotle was not particularly interested in prediction. Cf. Dicks (1970) 210f. on Aristotle's *Meteorology*, and evidence of his familiarity with the information found in the *parapēgmata*.

33 Aristotle *Prior Analytics* 1.1, 24b19–21; 1.4, 25b40–26a2. Aristotle discusses *sullogismos* in detail in the *Prior Analytics*. There he explains that a demonstration is a type of *sullogismos*, but not every *sullogismos* is a demonstration (25b30). Here I have used the definition in Barnes (2000) 51; see also Smith (1995) 30.

34 Freeland (1991) 50. See also Barnes (1975a) ix–xvii, and Barnes (1975b).

35 Aristotle *Posterior Analytics* 2.8, 93b9; cf. 2.10, 94a5–6. Trans. Barnes (*Complete Works*) 1:154.

36 Freeland (1991) 63. This is not the place to continue the debate. For a brief overview, interested readers should consult, e.g., Lloyd (1996b) 7–37, and Lennox (2001) 1–6.

37 Aristotle *Physics* 2.3, 194b23–35, trans. Hardie and Gaye (*Complete Works*) 1: 332–3.

There are extensive discussions in the philosophical literature about what Aristotle meant by these four causes. Barnes (1995b) 73 has argued that Aristotle 'seems to hold that there are four *types* or kinds of cause'. Hankinson (1995a) 120–1 has explained the four causes as 'four ways of answering different but equally crucial questions about why things are how they are'; they can be understood as four general classes of explanation. See Freeland (1991) 49–72 for an introduction to some parts of the debate. She takes the view that Aristotle understood the four *aitiai* (causes) to be four causes, and not explanatory factors, as some have suggested. See, e.g., van Fraassen (1980a) 131f. and (1980b) on this latter point, and also Fine (1987).

38 In the *Posterior Analytics*, Aristotle states that each of the four causes is 'proved' through the middle term (the '*B*' term in the example given above). The four types of causal relation (material, efficient, formal and final) are to be represented as 'middle terms' within the deductive syllogism. See Barnes (1975a) 214–21 on the challenges presented in interpreting this passage. See also Freeland (1991) 63 who points out that, within the formal syllogism, the different causes are each 'hidden' as the middle term, and it is not obvious which type of causation is being described.

39 Aristotle *Meteorology* 339a21–34.

40 Aristotle *Meteorology* 346b20–1.

41 For a statement of the importance of final causes, see Aristotle *Parts of Animals* 1.5, 645a24–7.

42 Aristotle *Generation of Animals* 5.1.

43 Aristotle *Physics* 2.8, 198b16–21; trans. Hardie and Gaye (*Complete Works*) 1: 339, with modification. See Furley (1985) 177; Sedley (1991); Wardy (1993).

44 This reading of the passage is not universally accepted. Some scholars see this passage differently, as Aristotle's presentation of a non-teleological process. See Furley (1985) 177 nn. 1 and 2, for references to both sides of the controversy. See also von Staden (1997) on teleology and mechanism.

45 See Sedley (1991) 185; Furley (1985) 178.

46 Aristotle *On Sleep* 457b31ff., trans. Beare (*Complete Works*) 1: 727.

47 Aristotle *Posterior Analytics* 96a2ff., trans. Barnes (*Complete Works*) 1: 158.

48 Aristotle *Physics* 198b34–199a8; trans. Hardie and Gaye (*Complete Works*) 1: 339. See also Furley (1985) 179; Sedley (1991); Wardy (1993).

49 See Judson (1991b).

50 Aristotle *Posterior Analytics* 1.30, 87b22–5; cf. 1.27, 43b31ff.

51 Aristotle *Metaphysics* 1027a20–6; trans. Ross (*Complete Works*) 2: 1622. See also Hankinson (1995a) 115–18. Scientific explanation is concerned with those things which happen always or for the most part; no *epistēmē* is possible of rare events. See Judson (1991b) 86.

52 Aristotle *Rhetoric* 2.22, 1396a1–3, trans. Rhys Roberts (*Complete Works*) 2: 2224.

53 In the *Posterior Analytics* 2.11, 94b32 Aristotle mentions a teleological cause of thunder (to threaten those in Hades in order to make them afraid) which he ascribes to the Pythagoreans.

54 See, e.g., Cooper (1982) particularly section 2, 202–16.

55 Sedley (1991) 195.

56 *On the Heavens* (e.g. 270b20–7) describes five elements (including the *aithēr*), while in *On Generation and Corruption* (e.g. 330a30) only four are present. That the *aithēr* is absent from the latter text is not surprising, since the work deals with coming-to-be and passing-away, and the *aithēr* is eternal (cf. *On the Heavens* 269a30–b17).

57 Aristotle *On the Heavens* 268b14–24.

58 Aristotle *Meteorology* 2.2, 339a11ff.; *On the Heavens* 1.2–3. As Geoffrey Lloyd has reminded me, it is important not to conflate the order of exposition with the chronological order of writing in Aristotle. Here, Aristotle's comment need not refer specifically to the writing of a treatise, but may simply refer to his philosophical work and discussions.

59 Aristotle *Meteorology* 339b16–30.

60 Cf. *On Generation and Corruption* 2.2–3, also *Meteorology* 4.1.

61 Aristotle *On the Heavens* 1.3 (269b23f.), and 4.1–4.

62 Aristotle *Meteorology* 340a13–17; cf. *On Generation and Corruption* 2.6, especially 333a16–27.

63 Aristotle *Meteorology* 339a36–b6; cf. *On Generation and Corruption* 2.4; *On the Heavens* 3.6 and 3.7.

64 Aristotle *On Generation and Corruption* 2.8.

65 Trans. Webster (*Complete Works*) 1:557. Lee (1952) 18–19 note *a* has pointed to the difficulty of understanding the exact location of this substance, noting that Ideler (1834–36) 1: 346 and Webster (1931) understood the passage to refer to 'the region between air properly so called and the moon', while he (following the extant ancient commentators on the *Meteorology*, as well as Heath (1913) 227–8, takes the passage as referring to the *aithēr*. The passage may, alternatively, refer to the 'fire-sphere' below the heavens, rather than the celestial region. See also Sharples (1990) 98–9. See also Lee (1952) 24–5 note *a* and Guthrie (1939) 178–9.

66 Trans. Webster (*Complete Works*) 1: 557. Aristotle then describes the substances in the terrestrial region (340b15–341a5f., trans. Lee (*LCL*) 19–21):

> The substance beneath the motion of the heavens is a kind of matter, having potentially the qualities hot, cold, wet and dry and any others consequent upon these: but it only actually acquires and has any of these in virtue of motion or rest. . . . So what is heaviest and coldest, that is, earth and water, separates off at the centre or round the centre: immediately round them are air and what we are accustomed to call fire, though it is not really fire: for fire is an excess of heat and a sort of boiling.

67 Trans. Lee (*LCL*) 21.

68 Aristotle *On the Heavens* 289a19–35.

69 It is interesting to note that he begins by suggesting that a full account of this would be more appropriate in a work on sensation, because heat is a sensible quality.

70 Trans. Lee (*LCL*) 25. There are problems in accounting for the sun's warming effect, as Freudenthal (1995) 104 has noted. Aristotle proposes two somewhat *ad hoc* hypotheses to account for this effect in *On the Heavens* 2.7 and *Meteorology* 1.3, 341b12ff. Freudenthal 104 note 70 mentions Longrigg's (1975) 214 view, that these explanations were 'both almost equally lame'. Additionally, there is also the problem of the sun's generative and vivifying effects; see Freudenthal (1995) 116.

71 Aristotle *Meteorology* 341b13–24; trans. Lee (*LCL*) 29–31. Aristotle here exhibits an interest in terminology and nomenclature evident elsewhere in

his work. Some other meteorological writers, notably Seneca, discussed in the following chapter, were also concerned with terminology.

72 Plato *Phaedrus* 270a; cf. *Cratylus* 396c.

73 The author of the Hippocratic treatise *On Ancient Medicine* (1.1) seems to distinguish between the *meteōra* and the things under the earth (ὑπὸ γῆν); trans. Jones (*LCL*) 1: 14.

74 Aristotle *Meteorology* 341b6ff., trans. Lee (*LCL*) 29. Commentators are not unanimous in their understanding of these exhalations. In particular, the nature of the exhalation that involves moisture (*atmis*) is understood by some to be hot, by others to be cool. See, e.g., Ross (1923) *Aristotle* 109–10; Lee (1952) 20–1; Hankinson (1995b) 153.

75 Aristotle *Meteorology* 359b28–34, trans. Lee (*LCL*) 165.

76 That there is no systematic discussion of the exhalations is not a criticism of Aristotle as an author. The *Meteorology* was probably not circulated as a polished philosophical work, but was very likely a working draft or lecture notes. The difficulties in understanding the characteristics of the exhalations, and the passing allusions to exhalation theory, may be explained by the supposition that the work is not a finished exposition.

Cf. Solmsen (1960) 406 who argues that Aristotle's meteorological theory completes his physical theory, integrating and connecting the various treatises, and Hall (1970) 324 who suggests that Aristotle was not writing on meteorology because of a particular interest in the subject, but to fill what would otherwise be a gap in his natural philosophy.

77 Cf. Aristotle *Meteorology* 341b33. See West (1978) 366–8 on some meteorological associations of 'goats'.

78 Aristotle *Meteorology* 342a16–27, 342a30–3, trans. Lee (*LCL*) 35.

79 Aristotle *Meteorology* 360a13ff., 360b30ff.

80 Aristotle *Meteorology* 361a31ff., trans. Lee (*LCL*) 173.

81 Aristotle *Meteorology* 361b1f. On the analogy between winds and rivers, see also Hall (1970) 299ff.

82 Aristotle *Meteorology* 368a10ff.

83 These are described only generally at the end of Book 3, but are discussed in greater detail in Book 4.

84 Aristotle *Meteorology* 338a20–339a9.

85 Cf. Lloyd (1996b) 37 *et passim*.

86 E.g. Aristotle *Parts of Animals* 653b34ff.; see also *History of Animals*, e.g. 486b17–21, 517a1f., 517b3ff.; *Generation of Animals* 730b5ff.

87 Aristotle *Metaphysics* 1026a6ff.

88 Aristotle *Posterior Analytics* 1.7 (75a38–9), trans. Barnes (*Complete Works*) 1: 122.

89 Aristotle *Posterior Analytics* 1.9 (78b35ff. and 79a14f.), trans. Barnes (*Complete Works*) 1: 128–9.

90 Aristotle *Posterior Analytics* 1.9 (79a12f.), trans. Barnes (*Complete Works*) 1: 129.

91 See Barnes (1975a) esp. 128–9 and 151–5.

92 Hankinson (1995a) 117.

93 I thank Marlein van Raalte for her help here.

94 Aristotle *Meteorology* 359a22–b4, trans. Lee (*LCL*) 161.

95 See Hankinson (1995b) 155–6 on Aristotle's 'false data'.

96 Frisinger (1977) 17.

97 Aristotle *Meteorology* 1.13, 349a26–7.

98 The translation of this term has been the focus of some debate. See Barnes (1980), and also Burnyeat (1982) 197 n. 11 on the importance of translating *endoxos* as 'reputable' and 'respectable'. On the 'factual' status of *endoxa*, see Pritzl (1994) 44; cf. Freeland (1990) 94 on 'reputable facts'.

99 Aristotle *Meteorology* 365a14ff.

100 See, e.g., Aristotle *Meteorology* 339b33. Freeland (1990) 70.

101 Aristotle *Meteorology* 348a14–30.

102 Aristotle's use of *endoxa* has been much discussed by modern philosophers and historians of philosophy. There is no agreement on the role of *endoxa* for Aristotle; this is, however, a bigger debate than can be addressed here. Freeland (1990) 76–8 provides a helpful discussion; at 78 she suggests that Aristotle 'regards it as important to see where an existing scientific theory fails, because it makes a false prediction about something that one might otherwise not have considered relevant, so might not otherwise have observed'.

103 See Aristotle *Topics* 100a20ff.; Lloyd (1996b) 218.

104 On stochastic explanations, see Taub (1997); see also Ierodiakonou (1995).

There is a work in the Aristotelian corpus, known as the *Problems*; it is unlikely that Aristotle was the author, although his name is attached to it. It is composed of thirty-eight books, which contain questions organized by topic. It is easy to imagine that the questions posed here might well reflect *endoxa*. A broad range of topics is covered, from problems connected with medicine, to problems associated with justice and injustice. Several of the topics are closely related to subjects covered in the *Meteorology*; for example, Book 23 is concerned with problems regarding salt water and the sea, Book 25 with problems related to the air and Book 26 with the winds.

105 Freeland (1990) 78.

106 Aristotle *Sophistical Refutations* 34 (183b15–184b8).

107 Freeland (1990) 94.

108 Aristotle *Prior Analytics* 70a8–10, trans. Freeland (1990) 83.

109 Aristotle introduces the terms 'sign' (*sēmeion*) and 'proof' or 'evidence' (*tekmērion*) into technical usage in the *Prior Analytics* (2.27), as part of his discussion of the terms used within inferences. Here, a *tekmērion* constitutes conclusive evidence for something. Although Aristotle does not mention it, the term is also used in legal contexts. A *sēmeion* is an indication or sign, which does not constitute conclusive evidence. Aristotle *Prior Analytics* 70b1–6; see also Freeland (1990) 83 and Smith (1989) 227. Hankinson (1995b) 154 n. 14 points out that the term *sēmeion* comes from forensic and medical contexts. In this sense, a sign points toward the truth, without entailing it.

110 See Burnyeat (1982) 199.

111 See Burnyeat (1982) 193–4 and also Allen (2001) 13–78, esp. 72–8.

112 Aristotle *Prior Analytics* 70a38; cf. Freeland (1990) 83, 92. See Burnyeat (1982) and also Freeland (1990) 83 *et passim*, for fuller discussions of signs

and inferences, and also the question of probability. See also Weidemann (1989), and Sedley (1982) on post-Aristotelian debates about signs.

113　Freeland (1990) 93. Further, as Burnyeat (1982) suggests and Freeland agrees, 'signs allow Aristotle the use of certain everyday forms of inference'; see Freeland (1990) 92ff. for a more detailed discussion. See also Lipton (1991).

114　Aristotle *Meteorology* 346b23–31, trans. Lee (*LCL*) 69–71.

115　Aristotle *Meteorology* 372b18–34, trans. Lee (*LCL*) 247–9.

116　See Freeland (1990) 90–4 for a discussion of this abductive argument.

117　Aristotle's use of analogy has itself been the subject of much scholarly study. See, e.g., Lloyd (1964) 77–8 and Lloyd (1966) particularly chap. 7, 'The Unity of Analogy'; Hesse (1966) chap. 7; Fiedler (1978). Freudenthal (1995) 117 n. 27 briefly mentions Aristotle's notion of analogy as functional equivalence, cf. e.g *Parts of Animals* 1.4 (644a17ff.); see also Balme (1972) 120, 148.

118　See Matthen (2001).

119　Aristotle *Meteorology* 366b14–19, trans. Lee (*LCL*) 209. On Aristotle's discussion of earthquakes, see also Hine (2002) 68–9 and 72–3.

120　Aristotle *Meteorology* 357b24–6, 358a3–25. Aristotle, a little further on in the text (358b16–18), refers to certain 'experiments' with liquids and condensation. See also Hankinson (1995b) 157.

121　Freudenthal (1995) 72–3 offers three possible interpretations of the nature of the earth's internal heat: (1) inherent, primeval heat; (2) derived from the sun; (3) primeval divine heat. He also suggests that Aristotle's 'occasional allusions to an analogy between the living animal body and the earth (e.g. *Meteorology* 1.14, 351a26f.) may be taken to lend some support' to the idea that the earth inherently possesses 'primeval' heat: 'Aristotle may have thought that the earth possesses an internal source of heat, in analogy with the heart, the source of vital heat in sanguineous animals'.

122　Aristotle *Meteorology* 341b6ff.; see also the discussion of the nature of the two exhalations above.

123　Aristotle *Meteorology* 2.4, 360a6ff., trans Lee (*LCL*) 165.

124　Cf. Eichholz (1949) on the role of the dry and hot exhalation to supply the heat which, *qua* efficient cause, forms minerals.

125　Aristotle *Parts of Animals* 2.3 (650a21–3), trans. Ogle (*Complete Works*) 1: 1012; cf. 4.4. (678a13). See also Freudenthal (1995) 71–3.

126　Aristotle *Parts of Animals* 2.3, 650a25. Aristotle *On Youth, Old Age, Life and Death, and Respiration* 1 (468a10ff.) draws an analogy between plant and animal digestion, stating that 'there is a correspondence between the roots in a plant and what is called the mouth in animals, by means of which they take in their food, some from the earth, some by their own efforts', trans. Ross (*Complete Works*) 1: 746. Cf. also the Hippocratic *Humours* 11, which states that 'just as earth is to trees, so is the stomach to animals'. See also Lloyd (1996b) 48.

127　Aristotle *Parts of Animals* 672b14ff.; according to Aristotle, in the case of humans, the diaphragm, that membrane which divides the region of the heart from the region of the stomach, serves to prevent the perceptive soul (located in the heart) from being overwhelmed by these bodily exhalations and heat. See also Lloyd (1996b) 48.

128 Aristotle *On Youth, Old Age, Life and Death, and Respiration* 470a20–b1, trans. Ross (*Complete Works*) 1: 749. See also Freudenthal (1995) 71.

129 Aristotle *Meteorology* 344a25ff., trans. Lee (*LCL*) 51–3.

130 Trans. Lee (*LCL*) 225.

131 Bourgey (1975).

132 Aristotle *Metaphysics* 9.6, 1048a35–7, trans. Ross (*Complete Works*) 2: 1655.

133 Hankinson (1995b) 155 suggests that Aristotle's analogies have no probative value; furthermore, he believes that Aristotle did not intend them to.

134 See, e.g., Aristotle *Meteorology* 2.3 (357a33ff.) on the analogy with the production of urine.

135 See Lloyd (1964) who also discusses the more general debate about the use of experiment in ancient Greek science, and von Staden (1975). Cf. Bourgey (1975); Lloyd (1987), chap. 5; Lloyd (1979) 222–5.

136 Aristotle *Meteorology* 358b34–359a14.

137 I am grateful to Keith Batchelor, of Blayson Olefines Ltd, Waterbeach, for telling me about the properties of wax, including raw beeswax. In the Fens, it is not unusual to wear a waxed jacket as rain protection.

138 As Lee has pointed out (1952) 158–9 note *a*, κεράμινον ('earthenware') has been suggested for κήρινον ('wax', Thompson (1910) note 3 on *Historia Animalium* 8.2, 590a22–7), but there is no manuscript support for this emendation. Both Pliny and Aelian have 'wax'. See also Lloyd (1964) 90–1 and Hankinson (1995b) 157–8; Bourgey (1975) 178 suggests that the wax acts as a filter.

139 Cf. Aristotle's reports about the evaporation of seawater and wine, *Meteorology* 358b16ff. See also Lloyd (1964) 90–1 regarding possible ways to replicate Aristotle's results. As he notes, 'if wine is evaporated and the vapour collected on a plate or some such object held over the boiling wine, the liquid which condenses is a colourless, almost flavourless fluid of low alcoholic content which would pass for "water" as naturally as the liquid collected from evaporating sea-water'; I thank Kemal de Soysa for trying the 'experiments' with me.

140 See Lee (1952) 67 note *b*; Jackson (1920); Netz (1999) 37 n. 65. See also note 76 above.

141 On lettered diagrams in Greek mathematics, see Netz (1999) 37.

142 See also Harley and Woodward (1987) 145.

143 Aristotle *Meteorology* 363a27–9, trans. Lee (*LCL*) 189.

144 Aristotle *Meteorology* 363a34–b7, trans. Lee (*LCL*) 189.

145 Aristotle *Meteorology* 363b8, trans. Lee (*LCL*) 189–91. A work known as 'The Situations and Names of Winds' is sometimes attributed to Aristotle. The manuscript version has the heading 'from Aristotle's treatise *On Signs*' and is probably a Peripatetic product. See Hett (1936) 451.

146 Thompson (1918) 53, pointing to the 'duodecimal or zodiacal division' of the diagram, suggests a Babylonian origin or influence.

147 See also Obrist (1997); I thank Bruce Eastwood for bringing this to my attention.

148 See Netz (1999) 36; cf. Aristotle *Meteorology* 375b18; *Prior Analytics* 41b14.

149 Cf. Aristotle *Posterior Analytics* 75b16, 76a24, 78b37 where optics is described as related to geometry and geometrical demonstrations apply to optical demonstrations; see also *Physics* 194a8. Aristotle discussed vision elsewhere, in *On the Soul* 2.7.2.

150 Aristotle *Meteorology* 371b18–21, 372a17–21; cf. 377b27ff.

151 Aristotle *Meteorology* 344a9–346b7.

152 Cf. the sharp distinction made in *On the Cosmos*, discussed in the next chapter.

153 Aristotle *Meteorology* 372a29ff.; *Posterior Analytics* trans. Barnes (*Complete Works*) 129.

154 Trans. Ross (*Complete Works*) 2: 1572.

155 Aristotle's views on the relationship between mathematics and the physical world have been much debated. Philosophers and historians have also grappled with Aristotle's philosophy of mathematics; a detailed consideration is not possible here. I have found Cleary (1995) especially helpful; he grounds his discussion of Aristotle's philosophy of mathematics within the contexts of the cosmological and metaphysical problems that led Aristotle to his views. See Cleary (1995) xxvff. for his explanation of his approach.

156 Hussey (1991b) 213.

157 Aristotle *Physics* 193b23–5, trans. Hardie and Gaye (*Complete Works*) 1: 331.

158 Aristotle *Physics* 194a8–10; cf. *Metaphysics* 997b15–21, 1078a14–18.

159 Aristotle *Physics* 194a10–11, trans. Hardie and Gaye (*Complete Works*) 1: 331.

160 Cleary (1995) 437–8.

161 Aristotle *Posterior Analytics* 79a2–6.

162 Aristotle *Meteorology* 372b12–17, trans. Lee (*LCL*) 247.

163 Aristotle *Meteorology* 373a6–16, trans. Lee (*LCL*) 249; Lee notes that 'here Aristotle in effect assumes what he is setting out to prove'.

164 Trans. Barnes (*Complete Works*) 128. See Aristotle *Physics* 194a8–10, also discussed above.

165 Aristotle *Meteorology* 373a17–19, trans. Lee (*LCL*) 249–51.

166 Aristotle *Meteorology* 373a32. For a more detailed discussion of Aristotle's treatment of the rainbow, see Heath (1949) 183–90; Boyer (1959) 36–54. See also Sorabji (1972).

167 Netz (1999) 212.

168 Netz (1999) 61.

169 Lloyd (1966) 426.

170 Netz (1999) 203, 210.

171 Netz (1999) 212.

172 It is very likely that Aristotle used diagrams elsewhere, for example in his anatomical, physiological and zoological studies. While, as in the *Meteorology*, the drawings themselves do not survive, at certain points in his

zoological works Aristotle seems to refer to diagrams. See Lloyd (1978) reprinted in Lloyd (1991) 231 n. 7, and also Kullmann (1998) 130.

173 Aristotle *Meteorology* 378a18ff., trans. Lee (*LCL*) 287, with modifications.

174 See also Eichholz (1949) 141–6, and Solmsen (1960) 401–3. The fourth book is not considered here.

175 Aristotle *Meteorology* 338b24f., trans. Lee (*LCL*) 5.

176 See Solmsen (1960) 399–402, 404–12; cf. Strohm (1935) 80ff., who argued that the exhalations did not play the same role throughout the *Meteorology*, and Strohm (1979) 181f. (note to 56, 10f. = 361b8f.).

177 Aristotle *Meteorology* 344a5ff., trans. Lee (*LCL*) 49.

178 See Cronin (1992) on the authorship of *On Weather Signs*. The author may not have been Theophrastus, but someone who used a work by him; see the previous chapter.

Diogenes Laertius, in his lengthy list of Theophrastus' writings, includes a Περὶ ἀνέμων, *On Winds*, in one book (at 5.42), Τῆς μεταρσιολεσ-χίας (*Metarsioleschia*), in one book (at 5.43), and Μεταρσιολογικῶν (*Metarsiology* = a dialect form of *Meteorology*), in two books (at 5.44); see Sharples (1998a) 16–18. The *Metarsioleschia* may have been a discussion of Democritus' ideas on meteorology; cf. Sharples (1998a) 27–8. It has been suggested that the *Meteorology*, *On Winds*, and *On Weather Signs* were all written as lecture notes, rather than as polished treatises. On the *Meteorology*, see Daiber (1992) 284; on *On Winds*, see Coutant and Eichenlaub (1975) xl; on *On Weather Signs* and *On Winds*, see Wood (1894) 10–11.

179 Diogenes Laertius 5.44. W. Capelle (1913) 333 suggests that Theophrastus may have chosen to use the term *metarsiology* to indicate that his own treatment would be restricted to 'atmospheric' phenomena. See also Regenbogen (1940) col. 1408 and Sharples (1998a) 17 n. 56. Theophrastus does discuss earthquakes. See also Strohm (1937) 407–11.

180 Daiber (1992) has provided a review of the texts that survive and has published the first full edition and English translation of the surviving Arabic version, apparently by Ibn al-Khammār, of the Theophrastean text; I have here relied on his translation. See also Drossaart-Lulofs (1955).

See Daiber (1992) 283–8 and Kidd (1992) on the question of completeness.

181 I use the form *Meteorology/Metarsiology* to refer to this work; modern scholars use both names for it; e.g. Mansfeld (1992) prefers *Metarsiology*, while many others, including Daiber (1992), opt for *Meteorology*.

182 Daiber (1992) 284–5; Mansfeld (1992); Sedley (1998a) 179–82. Sedley's suggestion is important in considering, in the next chapter, the influence of Theophrastus on later authors, and the texts to which they had access.

183 Daiber (1992) 285. Diogenes Laertius (5.44) mentions Theophrastus' *Metarsiology* in two books; Daiber suggests that the two books may have reflected the division he points to between above-the-earth and below-the-earth phenomena. Mansfeld (1992) 314–15 points out that we may

not have the full text and suggests that the section on earthquakes is not long enough to qualify as a 'book'.

Daiber (1992) 285 further argues that the *Meteorology* should be regarded as part of Theophrastus' Περὶ φυσικῶν; cf. Sharples (1998a) 2–4 and 7–8, who points to the ambiguity of the title, which could refer either to a doxographical work 'On the Natural Philosophers', or a physical work 'On Nature'. Daiber (1992) 286 suggests that the chronological order in which the works were written was Περὶ ὕδατος (*On Water*), μεταρσιολογικά, περὶ ἀνέμων, with Περὶ πυρός (*On Fire*) following.

184 On the absence of a discussion of comets in Theophrastus' *Meteorology*, see Steinmetz (1964) 216–17; see also Sharples (1998a) 153 and 160–1. On the Milky Way, see Sharples (1998a) 91 and 108–11.

185 Theophrastus *Meteorology/Metarsiology* [1] 1–23, [2], [3], [4], [5], [15] 2–16, in Daiber (1992) 261–71.

186 Furthermore, there were various attitudes among ancient thinkers towards multiple causation as an explanatory strategy. See I.G. Kidd (1992).

187 See Mansfeld (1999a) 17–19 on the use of the term 'doxography'.

188 Cf. Daiber (1992) 285.

189 Sharples (1998a) xv.

190 The adoption of multiple causation as an explanatory procedure characterized Epicurus' meteorology, discussed below. He seems to have regarded the various causes as possible, but not necessarily actually at work.

191 Sharples (1998a) xv.

192 Theophrastus *Meteorology/Metarsiology* [8] 2–4, Daiber (1992) 267.

193 Theophrastus *Meteorology/Metarsiology* [6] 22–5, Daiber (1992) 264.

194 Theophrastus *Meteorology/Metarsiology* [6] 29–67, Daiber (1992) 264–5.

195 Vallance (1988) 36 and 39 has also noted the relationship between Theophrastus' writings *De Lapidibus* (*On Stones*) and *De Igne* (*On Fire*) and the interest in 'problems' in later Greek science. Entire books of the Aristotelian *Problems* derive from Theophrastean works; cf. Flashar (1962) 320–2, 335–40. I am grateful to both Jim Hankinson and Bob Sharples for helpful comments here.

196 Theophrastus *Meteorology/Metarsiology* [I] 24–38, trans. Daiber (1992) 262.

197 Theophrastus *Meteorology/Metarsiology* [15] 26–7; trans. Daiber (1992) 271.

198 Theophrastus *Meteorology/Metarsiology* [1] – [2], Daiber (1992) 261–2. Kidd (1992) 300 has pointed out that the seven different causes of thunder are each related to seven types of noise: 'clap, grinding, rumbling, bursting sounds, slapping noise, "raspberry" (i.e. as of air escaping from a skin or balloon), hissing crackle'. As each of the sounds is different, 'the physical conditions or causes must be different, and so they *are* given differently'.

199 Theophrastus *Meteorology/Metarsiology* [9] 2ff. is an account of the causes of snow. For whiteness in particular see [9] 8–11, Daiber (1992) 267.

In *On Winds* 13 Theophrastus mentions briefly the claim by 'the Cretans, among others' that winters in earlier times were more severe, with more snow. Here, he seems to accept the view that weather patterns change over time. Or he may simply be using a well-rehearsed *topos* regarding the greater severity of winters past. In the *Meteorology* (351a19ff.), Aristotle considered the silting of rivers to represent long-term change.

200 Theophrastus *Meteorology/Metarsiology* [13] 3–5; Daiber (1992) 268. See also Sharples (1998a) 147–8 and n. 419 on the exhalations and the formation of wind in Theophrastus.

201 Theophrastus *Meteorology/Metarsiology* [13] 14–18; Daiber (1992) 268. See also Vallance (1988) 38.

202 Theophrastus *Meteorology/Metarsiology* [13] 43f., trans. Daiber (1992) 269; see Daiber (1992) 279, 13.33–42 for note on '*WRS*'.

203 Theophrastus *Meteorology/Metarsiology* [13] 33–42; Daiber (1992) 269.

204 Theophrastus *On Winds* 1, trans. Coutant and Eichenlaub (1975) 3.

205 See Daiber (1992) 286.

206 Theophrastus *On Winds* 1, trans. Coutant and Eichenlaub (1975) 3.

207 Theophrastus *Meteorology/Metarsiology* [13] 8–17; Daiber (1992) 268; cf. *On Winds* 10. See also Sharples (1998a) 150–1.

208 Theophrastus *On Winds* 15, trans. Coutant and Eichenlaub (1975) 17.

209 Theophrastus *On Winds* 17. Woods (1894) 30 n. 24 is helpful concerning the reading of κατὰ σύμπτωμα here. He discusses Schneider's translation as 'fortuito', i.e. 'by chance', in contrast with his own translation 'according to a regular concurrence'. Coutant and Eichenlaub (1975) 17 translate the phrase somewhat similarly, as 'in conjunction'. This reading seems to be consistent with Theophrastus' approach in the rest of the text.

210 Cf. Kidd (1992) 303f.

211 The relationship of Theophrastus' meteorology to that of Aristotle has been addressed by several scholars; Sharples (1998a) 146ff. See also Sharples (1998b).

212 Sharples (1996b); Sharples (1998b) 279. Theophrastus' views on teleology, particularly as they relate to those of Aristotle, have also been debated; see, e.g., Lennox (1985) and Repici (1990).

213 Theophrastus *Metaphysics* 8b24–9a9, trans. Ross and Fobes (1929) 25. See also the useful comments concerning Theophrastus' use of analogy, in van Raalte (1993) 427–8.

214 Theophrastus *Metaphysics* 9a18ff., trans. Ross and Fobes (1929) 27.

215 Theophrastus *Metaphysics* 9b22ff., trans. Ross and Fobes (1929) 29.

216 See, e.g., *On Winds* 14, 17, 34 and 42.

217 Theophrastus *Metaphysics* 9a10ff. See also Laks and Most (1993) 67–8.

218 See Vallance (1988) 36–9.

219 On the exhalations, cf. Kidd (1992) 295. See also Sharples (1998a) 147f. and Janković (2000) 17–19. There is considerable debate regarding the relationship of Theophrastus' exhalations to those of Aristotle. See Sharples (1998a) 146–52 for a useful summary.

220 See Kidd (1992) 295 who suggests that the order in Theophrastus' *Meteorology* may be elemental: Fire, Water, Air and Earth. Some scholars

emphasize the role of the element Fire in Theophrastus' physics, but it is not entirely clear that Fire plays a special role in his *Meteorology*; Vallance (1988) 35–6 argues against this view. See also Sharples (1998a) 89f. and 113–16.

221 See also Sharples (1998a) 149–51.
222 Cf. Wöhrle (1985) 129–38.
223 See Vallance (1988) 36.

4 METEOROLOGY AS A MEANS TO AN END: PHILOSOPHERS AND POETS

1 Sharples (1998a) 145–6 provides a useful guide, with bibliographical references, to the use of Theophrastus' explanations by later writers.
2 Theophrastus *Meteorology/Metarsiology* 14.14–29, trans. Daiber (1992) 270, who has indicated by his marks that 'thunderbolt' appears in the original in the plural form. Additions to the English translation not included in the original text are indicated by '( )' and '< >'; additions by the Arabic/Syriac translator/transmitter by '{ }'.
3 Daiber (1992) 280.
4 Van Raalte (forthcoming) argues against Theophrastus' authorship of this passage; I am grateful to her for sharing a pre-publication version. That the gods do not employ meteorological phenomena as a means of punishment does not mean that there is no divine element in Theophrastus' world-view. See also Mansfeld (1992b) 314–35; Hankinson (1998) 191–3.
  It is possible that Theophrastus held different views at different times in his life. For ancient accounts of his theological views, see FHS&G 1: 442–55, fr. 251–63. For Theophrastus' views regarding the place of the divine with regard to the motion of the heavens, and the nature of his general view of the world as a system, see his *Metaphysics*, and FHS&G 1: 319ff., fr. 158ff.
5 See Mansfeld (1992b) 320–2.
6 See Mansfeld (1992b) and David Sedley (1998a).
7 I adopt the term 'didactic' letter from Mansfeld (1999a) 5.
8 Epicurus 'Letter to Pythocles', in Diogenes Laertius, *Lives of Eminent Philosophers* 10.83–117, trans. Hicks (*LCL*) vol. 2: 612–43. Here, Diogenes Laertius 10.85, trans. Hicks 2: 615, with slight alterations. (I have generally used Hicks' widely available translation; occasionally, I have chosen another, for clarity. I translate *ta meteōra* as 'meteorological', rather than 'celestial', phenomena.)
  In the 'Letter to Pythocles' Epicurus also made brief reference to his larger work, *On Nature*. See Diogenes Laertius 10.91–2.
9 Diogenes Laertius 10.115–16, trans. Hicks (*LCL*) 2: 641–3.
10 The title *De rerum natura* is translated in various ways. The range of scholarship on Lucretius is vast. Interested readers are directed, as a beginning, to Fowler and Fowler (1997) ix–xxx, as well as the 'Select Bibliography', xxxi–xxxiv.
11 On Epicurean theology, see Long and Sedley (1987) vol. 1: 63–4, 139–44 and 144–9, and also J. Mansfeld (1999b) esp. 454ff., 463f., 478.

12 Mansfeld (1992b) 326–7; Sedley (1998a) 157–60. See also Sedley (1998b) 354. Van Raalte (forthcoming), while rejecting Theophrastus' authorship of the 'excursus', has suggested that it might have grown out of his lecture notes.

13 Lucretius 6.388–422, trans. Melville (1997) 190–1.

14 Diogenes Laertius 10.86 and 10.94, trans. Hicks (*LCL*) 2: 615 and 623.

15 Diogenes Laertius 10.87, trans. Hicks (*LCL*) 2: 617.

16 See Kidd (1992) 294–306, particularly 303–4, and Allen (2001) 197.

17 Diogenes Laertius 10.86.

18 Diogenes Laertius 10.97, trans. Hicks (*LCL*) 2: 625.

19 Diogenes Laertius 10.97, trans. Hicks (*LCL*) 2: 625.

20 Diogenes Laertius 10.113–14, trans. Hicks (*LCL*) 2: 641.

21 Diogenes Laertius 10.98–9.

22 Diogenes Laertius 10. 79–80, trans. Hicks (*LCL*) 2: 607–9.

23 Diogenes Laertius 10.85–6, trans. Hicks (*LCL*) 2: 615.

24 Diogenes Laertius 10.86–7, trans. Hicks (*LCL*) 2: 629, with some alteration.

25 Diogenes Laertius 10.80 ('Letter to Herodotus'), trans. Asmis (1984) 322.

26 Diogenes Laertius 10.87, trans. Hicks (*LCL*) 2: 617.

27 E.g. Diogenes Laertius 10.80.

28 Diogenes Laertius 10.100, trans. Hicks (*LCL*) 2: 627–9.

29 Diogenes Laertius 10.102, trans. Hicks (*LCL*) 2: 629, with slight alteration (i.e. translating *phainomena* as phenomena, rather than 'facts').

30 Diogenes Laertius 10.104, trans. Hicks (*LCL*) 2: 631.

31 Diogenes Laertius 10.109, trans. Hicks (*LCL*) 2: 637; Lucretius 6.527ff., trans. Latham (1951) 233.

32 Sedley (1998a), particularly 125–6, 145–6, 157, 182–5. See also Mansfeld (1992b), and Asmis (1984) 328–9. Epicurus did not agree with everything that Theophrastus had to say; Epicurus also wrote a work *Against Theophrastus*.

33 I am grateful to Geoffrey Lloyd for having helped me think through the differences here. See also Hankinson (1998) 221–3; Allen (2001) 197.

34 Kidd (1992) 305 has drawn attention to the specific order in which various meteorological phenomena are considered by Aristotle, Theophrastus, Epicurus, Lucretius and Seneca.

35 See Mansfeld (1992b). While comets were not treated by Theophrastus in the extant parts of the *Meteorology/Metarsiology*, he did discuss them elsewhere: see FSH&G 160–1, fr. 193, and Sharples (1998a) 160–1.

36 Toohey (1996) 95–6. The poem is unusual in its length; most ancient didactic epics consisted of only a single book. It is not clear whether the poem was completed. See Fowler and Fowler (1997) 'Introduction' xxii, and also Sedley (1998a) 160–5.

37 Lucretius 6.39–41, trans. Melville (1997) 180.

38 Lucretius offers explanations for thunder (6.96–159), lightning (6.160–218), thunderbolts (6.219–422), waterspouts (sg. = *prēstēr*) (6.423–50), earthquakes (6.535–607), volcanoes (6.639–702, including a description of Aetna), the flooding of the Nile (6.712–37) and magnets (6.906–1089).

On the meaning of the term *prēstēr* (πρηστήρ), see note 28 to Chapter 3. Sedley (1998a) 157–60 points to Epicurus' *On Nature* (Book 13) as Lucretius' source; cf. also 135–44. See Fowler and Fowler (1997) xxiv–xxv for a brief discussion of the possible sources of Lucretius' poem.

39 Lucretius 6.703–11, trans. Melville (1997) 200. Asmis (1984) 324 offers an alternative (prose) translation:

> There are also some things for which it is not enough to state a single cause, but (one must state) several, of which one, however, is the case. Just as if you were to see the lifeless corpse of a man lying far away, it would be fitting to state all the causes of death in order that the single cause of this death may be stated. For you would not be able to establish conclusively that he died by the sword or of cold or of illness or perhaps by poison, but we know that there is something of this kind that happened to him. We can say the same thing in many cases.

40 Asmis (1984) 324.
41 I thank David Sedley for reminding me of this contrast.
42 Diogenes Laertius 10.100.
43 Lucretius 6.129–31. See also Asmis (1984) 327.
44 Lucretius 6.145–9, trans. Melville (1997) 183.
45 Lucretius 6.165–72, trans. Melville (1997) 184.
46 Trans. Sedley (1998a) 35.
47 See, e.g., Cicero *On the Nature of the Gods* 2.22–3, 28–30; Diogenes Laertius 7.138–9. See also Sambursky (1959) 21–44.
48 Diogenes Laertius 7.151–4.
49 Nothing is known about Manilius' life. Gould (1977) xii dates Books 1 and 2 to the period when Augustus (who died in 14 CE) was still alive; Books 4 and 5 were written after his death.
50 Toohey (1996) 180–4 argues that Manilius was writing in direct opposition to Lucretius.
51 Manilius *Astronomica* 2.60–76, trans. [Creech] (1697) 51–2. I have chosen this translation to give a sense of the poetic form of the work. I here append a more modern translation; further translations used here of the *Astronomica* are by Goold (*LCL*).

Manilius *Astronomica* 2.60–76, trans. Goold (*LCL*) 87–9:

> For I shall sing of God, silent-minded monarch of nature, who, permeating sky and land and sea, controls with uniform compact the mighty structure; how the entire universe is alive in the mutual concord of its elements and is driven by the pulse of reason, since a single spirit dwells in all its parts and, speeding through all things, nourishes the world and shapes it like a living creature. Indeed, unless the whole frame stood fast, composed of kindred limbs and obedient to an overlord, unless providence directed the vast resources of the skies, the earth would not possess its stability, nor stars their orbits, and the heavens would wander aimlessly or stiffen with inertia; the constellations would

not keep their appointed courses nor would alternately the night
flee day and put in turn the day to flight, nor would the rains feed
the earth, the winds the upper air, the sea the laden clouds, rivers
the sea and the deep the springs; the sum of things would not
remain for ever equal through all its parts.

52 Aristotle *Meteorology* 2.4, 360a6–18; Aristotle *On Youth, Old Age, Life and Death, and Respiration* 470a20–b1. But see also Matthen (2001) on the limits of understanding the Aristotelian cosmos as a living creature.

53 But while Epicurus discussed comets, Lucretius did not; comets are not treated in extant parts of Theophrastus' *Meteorology*.

54 Trans. Goold (*LCL*) 71.

55 Manilius *Astronomica* 1.817–66: earth-vapours ignited by dry air; 1.817–26 quoted here, trans. Goold (*LCL*) 71.

56 Manilius *Astronomica* 1.831f., trans. Goold (*LCL*) 71. As Goold xxxvi notes, 'Manilius's types of comet are readily identifiable with the classifications we meet with elsewhere (*e.g.* De Mundo 4; Seneca *NQ* VII *passim* and I.15.4; Pliny, *Natural History* 2.89 ff.; cf. Ptolemy, *Tetrabiblos* 2.9).'

57 Manilius *Astronomica* 1.853–8, 1.859–64, trans. Goold (*LCL*) 73.

58 Manilius *Astronomica* 1.893–7, trans. Goold (*LCL*) 75–7.

59 It is questioned whether or not Seneca should properly be classified as a 'Stoic'. On Seneca and his philosophical orientation, see Maurach (1975a); Griffin (1976/92); Inwood (1995).

60 Seneca (*NQ* VI.1) refers to the Campanian earthquake of 62 or 63; see Griffin (1976/92) 399–400 Note H on questions regarding the dating of the earthquake. Senecan scholars do not agree how much of the *NQ* survives. See, e.g., Gross (1989); Hine (1981); Waiblinger (1977); Brok (1995).

61 See Hine (1981) 6–19 and (1996) xxii–xxv, who adopted the order transmitted by the manuscripts. His ordering has been accepted by other scholars (notably Codoñer (1989) and Inwood (1999, 2002)), and rejected by Gross (1989). Hine has suggested the following order: Book 1 (= Book III of the 'traditional' order) on rivers, Book 2 (= IVA) on the Nile, Book 3 (= IVB) on <clouds, rain> snow, hail, Book 4 (= V) on winds, Book 5 (= VI) on earthquakes, Book 6 (= VII) on comets, Book 7 (= I) on meteors, rainbows and related phenomena ('fires in the air') and Book 8 (= II) on thunder and lightning. For the convenience of readers, and to minimize confusion in citations, the traditional book numbers (using Roman numerals) will be used here, even though the order of discussion will, for the most part, follow Hine's suggestions.

62 See Inwood (1999) and (2002). I am grateful to Professor Inwood for having shared a pre-publication version with me.

63 See Corcoran (1971/72) 1: xii; Goodyear (1984). See also Hine (2002) 60–3.

64 The title of the *Natural Questions* is somewhat problematic, for the structure of the work is unlike that of other examples of 'problem' (*problēmata*) literature. Hine (1981) 24–9 discussed some of the difficulties involved in determining the original title.

65 What we know of Apollonius is information provided by Seneca. Pliny (7.193) refers to Epigenes as an 'authority of the first rank'. See Corcoran (1971/72) 2: 233 n. 4, and Hall (1977).

Only fragments survive of the writings of Posidonius of Apamea (c. 135 to c. 51–50 BCE), whose ideas on meteorology were widely cited by later ancient authors. Consequently, modern scholars have used descriptions of Posidonius' ideas by other ancient authors to try to reconstruct his views. This task is not straightforward, as it is difficult to ascertain which fragments are rightly attributed to Posidonius and various controversies have ensued regarding the reconstruction of his ideas and his influence on other writers. As Clarke (1999) 130 has noted, 'there is a curious disparity between the tiny fraction of his work to survive and the great reputation which has become attached to him'. In the assessment of Kidd (1996), the editor of the invaluable volumes of Posidonius' fragments, Posidonius was 'a dominant figure in his lifetime, [whose] subsequent reputation and influence have been overstressed to pandemic proportions and require re-examination; but the impact was considerable and continued at least to the 6th cent. AD'; see also Kidd (1988).

Those fragments that name a particular work written by Posidonius do not contain much specific information about his meteorological ideas, other than the titles of his books; Diogenes Laertius gives two titles of works by Posidonius on meteorology, an *Elementary Treatise on Meteorology* (7.138) and *About Meteorological Phenomena* (7.135). Those fragments that do provide more detail about his meteorological ideas are not attributed to any specific writing of his (see Edelstein and Kidd (1972) 122–33, Fragments 131–8, trans. Kidd (1999) 184–98. These few fragments give some information about his theory of comets, haloes, the rainbow, thunderstorms and lightning, hail, the classification of winds, hurricanes and earthquakes, volcanoes (Mt Aetna) and tides. The sources for these fragments vary (and in some cases are problematic). It is difficult to discuss Posidonius' meteorological views with any confidence, and such a speculative project is unlikely to add a great deal to our understanding.

Much of what we think we know about Posidonius' meteorological ideas comes to us through reports contained in Seneca's *Natural Questions.* Seneca's role in transmitting the views of Posidonius was crucial, for in many cases he is the sole source for ideas about various meteorological phenomena that he attributed to Posidonius.

66 The original format of the *NQ* is not entirely clear. Some surviving books (IVA and IVB) are incomplete. Furthermore, it may be the case that some books are missing in their entirety; there may, originally, have been separate books dealing with the sea and with volcanoes. See Hine (1981) 34.

67 As explained above, the order suggested by Hine is that used here. The first book, in his view, is Book III in the traditional ordering.

68 Seneca *NQ* III.Pref. 1 and 6–7, trans. Corcoran (*LCL*) 1: 201, 205.

69 Seneca *NQ* III.Pref.18.

70 Aristotle and Lucretius had also discussed the subject of rivers. In Book III (III.1.2), Seneca explains that he will discuss the Nile separately; it is the subject of Book IVA.

71 Seneca *NQ* III.10.1, III.10.4, trans. Corcoran (*LCL*) 1: 223–5.
72 Cf. Theophrastus *Meteorology/Metarsiology* 13.8–17, Daiber (1992) 268.
73 Seneca *NQ* III.15.1–6, trans. Corcoran (*LCL*) 1: 233–5.
74 Seneca *NQ* III.22.2, trans. Corcoran (*LCL*) 1: 255. In the course of Book III, and the *Natural Questions* more generally, Seneca provides interesting details of Roman life; there are other bathroom analogies in the work, e.g. at *NQ* IVB9.1.
75 Seneca *NQ* IVA.2.2.
76 Among philosophers cited in Book IVA are Thales (IVA.2.22) and Theophrastus (IVA.2.16). Seneca was also very interested in and influenced by poetry. He frequently quotes Virgil and Ovid (and even occasionally Lucretius); his use of Latin epic throughout the *Natural Questions* is striking.
77 Seneca *NQ* IVA.2.17; cf. Aeschylus fr. 300N = fr. 161 (*LCL* 2: 478–9) and *Suppliant Maidens* 561; Euripides *Helen*, 1 sqq. Herodotus (2.22) rejected the explanation that the runoff of melted snow from Ethiopian mountains feeds into the Nile. See Corcoran (*LCL*) 2: 32–3.
78 Seneca *NQ* IVB3.6–4.1, trans. Corcoran (*LCL*) 2: 51.
79 Seneca *NQ* IVB5.1, trans. Corcoran (*LCL*) 2: 53.
80 Seneca *NQ* IVB6.1, trans. Corcoran (*LCL*) 2: 55.
81 Seneca *NQ* IVB6.2, trans. Corcoran (*LCL*) 2: 55. For another ancient reference to weather-gear, see Diogenes Laertius 2.10, on Anaxagoras' rain-wear.
82 Seneca *NQ* IVB6.3, trans. Corcoran (*LCL*) 2: 55.
83 Seneca *NQ* IVB7.2–3, trans. Corcoran (*LCL*) 2: 57.
84 Seneca *NQ* IVB5.2–3, trans. Corcoran (*LCL*) 2: 53. When he describes a view he credits to Democritus, that 'the more solid an object is the more quickly it absorbs heat and the longer it retains it', Seneca adds that 'if you place a bronze vase and a glass vase in the sun, the heat will communicate itself more quickly to the bronze vase and cling to it longer'. (Seneca *NQ* IVB9.1, trans. Corcoran (*LCL*) 2: 59.) It is not clear whether Democritus based his idea on experience with vases; glass vases were more common in Seneca's time than in that of Democritus.
85 Seneca *NQ* III.7.1, trans. Corcoran (*LCL*) 1: 219.
86 In the context of meteorology, cf. Aristotle's discussion of the difficulty of describing the exhalations (*Meteorology* 359b28–33, and also 341b13ff.), and Theophrastus on the *prēstēr* (Theophrastus *Meteorology/Metarsiology* [13] 43–4; Daiber (1992) 269).
87 Seneca *NQ* V.2.1, trans. Corcoran (*LCL*) 2: 79.
88 Seneca *NQ* V.4.1–2, trans. Corcoran (*LCL*) 2: 81.
89 Elsewhere in Book V, Seneca points to similarities between water and air, e.g. at V.13.1–2 both water and air can form into swirling phenomena (whirlpools and whirlwinds); at V.14.2 Seneca notes that both can exist underground.
90 This interest in language and terminology appears at several points in the *Natural Questions*; see, e.g., Seneca *NQ* V.11.1, V.12.1.
91 Surviving school grammars from this period specifically comment on the examples of 'defective' nouns, which don't have examples of all six cases; Ineke Sluiter, personal communication.

92 Seneca *NQ* V.17.5, trans. Corcoran (*LCL*) 2: 113.
93 Cf., e.g., Theophrastus *On Winds* 2–9; pseudo-Aristotle *The Situations and Names of Winds*.
94 Cf. Inwood (2002) 137.
95 Seneca *NQ* VI.3.4, trans. Corcoran (*LCL*) 2: 143. See Griffin (1976/92) 399–400 Note H on questions regarding the dating of the earthquake.
96 Seneca *NQ* VI.4.2, trans. Corcoran (*LCL*) 2: 145. See also Hine (2002) 65–6 on Seneca's descriptions of specific earthquakes.
97 Seneca *NQ* VI.6.1–9.3.
98 Seneca *NQ* VI.12.1, trans. Corcoran (*LCL*) 2: 163.
99 Seneca *NQ* 6.20 = DK 68A98. As Asmis (1984) 329 has pointed out, that Aristotle (*Meteorology* 365b1–6 = DK68A97) attributed only the first of these explanations to Democritus may indicate that Aristotle, for his own reasons, chose to present only one of his explanations.
100 Seneca *NQ* VI.19.1–2, trans. Corcoran (*LCL*) 2: 183.
101 Seneca *NQ* VI.14.1–2, trans. Corcoran (*LCL*) 2: 169.
102 Seneca *NQ* VI.16.1–3.
103 Seneca *NQ* VI.24.2–3, trans. Corcoran (*LCL*) 2: 195–7.
104 Seneca *NQ* VI.17.1–2.
105 Seneca *NQ* VI.18.6–7, trans. Corcoran (*LCL*) 2: 181–3. Seneca also compares moving air and a flowing river. When a river is allowed to flow smoothly, everything is fine, but when the flow is impeded or held back, the river rages at the obstruction and causes damage.
106 Seneca *NQ* VI.32.1–2, trans. Corcoran (*LCL*) 2: 215.
107 See, e.g., Seneca *NQ* VII.12.4–7.
108 Seneca *NQ* VII.11.2, trans. Corcoran (*LCL*) 2:249. Aristotle (1.7, 344a23–4) described only two types of comets; Pliny (2.89), with reference to the views of the Greeks, mentioned only these two.
109 Seneca *NQ* VII.22.1–23.3; trans. Corcoran (*LCL*) 2: 275.
110 Seneca *NQ* VII.27.6, trans. Corcoran (*LCL*) 2: 285.
111 Seneca *NQ* VII.28.1–3, trans. Corcoran (*LCL*) 2: 287–9. On comets as signs of non-meteorological events, see Barton (1994b) 39–40; Gee (2000) 40–1, 154–5, 157ff.
112 Seneca *NQ* I.1.1, trans. Corcoran (*LCL*) 1: 15.
113 Seneca *NQ* I.3.11 and I.3.14, trans. Corcoran (*LCL*) 1: 41.
114 Seneca *NQ* 1.5.13, trans. Kidd (1999) 192.
115 Seneca *NQ* I.4.1, trans. Corcoran (*LCL*) 1:43; *Epistles (Letters to Lucilius)* 88: 10–13.
116 Seneca *NQ* I.5.14, trans. Corcoran (*LCL*) 1: 55.
117 See, e.g., *NQ* I.13.3.
118 Seneca *NQ* II.1.3, trans. Corcoran (*LCL*) 1: 101.
119 Seneca *NQ* II.11.3.
120 Seneca *NQ* II.4.1, trans. Corcoran (*LCL*) 1: 105.
121 Seneca *NQ* II.4.1, trans. Corcoran (*LCL*) 1: 105.
122 Seneca *NQ* II.6.1, trans. Corcoran (*LCL*) 1: 107. See also Seneca *NQ* II.5.1–2.
123 Seneca *NQ* II.6.1, trans. Corcoran (*LCL*) 1: 107.
124 Seneca *NQ* II.10.2–3, trans. Corcoran (*LCL*) 1: 115–17.
125 Seneca *NQ* II.11.1, trans. Corcoran (*LCL*) 1: 117.

126 Seneca *NQ* II.11.2, trans. Corcoran (*LCL*) 1: 117–19.

127 Seneca *NQ* II.14.2, trans. Corcoran (*LCL*) 1: 127, with slight alterations.

128 Seneca *NQ* II.12.1, trans. Corcoran (*LCL*) 1: 119.

129 Seneca *NQ* II.12.2, trans. Corcoran (*LCL*) 1: 119.

130 This reference to coins is reminiscent of the passage in Theophrastus' *Meteorology/Metarsiology* 6.85–91; Daiber (1992) 265f. See Mansfeld (1992b) 316–17 on the passage in Theophrastus and whether there has been a dislocation involving the loss of other material.

131 Seneca *NQ* II.31.1–2, II.32.

132 Seneca *NQ* II.32.2, trans. Corcoran (*LCL*) 1: 151.

133 I thank Bob Sharples for helping me to clarify this point.

134 Seneca *NQ* II.33.1, trans. Corcoran (*LCL*) 1: 155.

135 Seneca *NQ* II.38.4.

136 Corcoran (*LCL*) 1: 163 n. 3.

137 Seneca *NQ* II.46.1, trans. Corcoran (*LCL*) 1: 175.

138 Seneca *NQ* II.45.1, trans. Corcoran (*LCL*) 1: 173.

139 Seneca *NQ* II.42.3, trans. Corcoran (*LCL*) 1: 171. Cf. Sextus Empiricus *Against the Professors* (*Adversus mathematicos*) 9.54 (= *Against the Physicists* 1.54; *LCL* 3: 28–33) on one of the fifth-century tyrants Critias' account of the ancient lawgivers' invention of the gods in order to inspire fear in humans. Modern scholars have credited the fragment quoted by Sextus to Euripides' satyr play *Sisyphus*, rather than Critias; see Kahn (1997) and Graver (1999). I am grateful to Marlein van Raalte for directing me to the Sisyphus fragment.

140 Seneca *NQ* II.45.1–3, trans. Corcoran (*LCL*) 1: 173.

141 Attalus was presumably a Stoic; Seneca refers to him (II.50.1) as 'our Attalus' (*Attalus noster*), although this may be a reference to Attalus as a teacher and philosopher from whom Seneca learned. He may have been a friend of Seneca; see Hine (1981) 410f. and Mansfeld (1992b) 330 n. 45.

142 Seneca *NQ* II.59.1–3, trans. Corcoran (*LCL*) 1: 193.

143 Inwood (2002) 156; cf. Inwood (1999) 43.

144 Hine (1981) 40 has noted, 'the artistic structuring of [the] whole book is in keeping with the literary qualities of the *NQ*, qualities which, like the linking of the scientific to the ethical, make the work closer in spirit to Lucretius than to Aristotle'.

145 Seneca *NQ* VII.32.4, trans. Corcoran (*LCL*) 2: 297–9. Cf. Inwood (2002) 147f. This sort of complaint was not confined to Seneca; rather, it formed the basis of a rather widespread *topos*. Cf. Pliny, below.

146 Seneca *NQ* I.Pref.14, trans. Corcoran (*LCL*) 1: 11–13.

147 Seneca *NQ* I.Pref.16–17, trans. Corcoran (*LCL*) 1: 14.

148 Inwood (2002) 156.

149 Seneca *To Helvia on Consolation* 20.1–2, trans. Basore (*LCL*) 2: 487–9. Seneca was in exile from 41 to 49 CE.

150 See Reale and Bos (1995) who maintain that *On the Cosmos* is an authentic work of Aristotle; see also Schenkeveld (1991) and Barnes (1977). On dating, see Furley (1955) 337–41. While the author remains unknown, the work shows links to the ideas of several philosophical schools.

The influence of Peripatetic and Neo-Pythagorean writers, as well as Posidonius, has been traced by various scholars. See Capelle (1905) and Maguire (1939) as well as Furley (1955).

151 See Furley (1955) 334–5.
152 Diogenes Laertius 10.97, trans. Hicks (*LCL*) 2: 625. See Long and Sedley (1987) 1: 63–4, 139–44 and 144–9 for an important discussion arguing that Epicurean gods are primarily moral concepts, rather than extraterrestrial living beings.
153 θεολογῶμεν = 'let us theologize'.
154 *On the Cosmos* 392b3–5, 392b8–13, trans. Furley (*LCL*) 353–5.
155 The author of *On the Cosmos* employs various analogies with human activities. For example, at the beginning of chapter 5 (396a33–b5), he notes that some people have questioned how the cosmos has survived, if indeed it is composed of opposites (namely, dry and wet, cold and hot). The author suggests that such survival is similar to the manner in which a city survives when it is composed of various types of people, rich and poor, old and young, weak and strong, good and bad. In his description (399b1ff.) of the way in which the god gives a sign to every moving thing to move, and then things move, he invokes an analogy with the trumpet signal given in a military camp, which stirs all members of the force in their individual tasks towards readiness for war. Other analogies with human experience occur as well (cf. the description of the god as a keystone, and the analogy to the sculptor Pheidias, 399b29ff.).
156 See also Strohm (1953) esp. 278–87, and Strohm (1979) 295–323.
157 *On the Cosmos* 394a23, 394a32, 394b2.
158 *On the Cosmos* 395a29ff. This passage has been the focus of debate regarding the influence of Posidonius on the author of *On the Cosmos*. The debate is not helpful for our purposes, as the scholars line up according to their views on Posidonius' influence more generally. Maguire (1939) 129 describes this as a somewhat 'peculiar' passage; he suggests that it was written as an afterthought, but believes that it showed Posidonian influence; this is the only part of the chapter on meteorology which Maguire thinks did show the influence of Posidonius, but he doesn't discuss this in detail. On the relationship of *On the Cosmos* to Posidonius, see Capelle (1905) 529–68, esp. 547–50; Strohm (1979) 312.
159 In the *Meteorology*, Aristotle does not use σέλα (lights) or διάττοντες (shooting stars), words found in *On the Cosmos*.
160 Theophrastus *Meteorology/Metarsiology* [14]; Daiber (1992) 269.
161 Trans. Furley (*LCL*) 377. See de Boer *et al.* (2001) on new evidence for the geological origins of the ancient Delphic oracle.
162 Maguire (1939) 131 has argued that the source here may have been Asclepiodotus (who is cited by Seneca in connection with earthquakes, *NQ* 6.22). On Asclepiodotus, who may have been a student of Posidonius, see Illinois Greek Club (1923) 230–8.
163 *On the Cosmos* 397a8ff., trans. Furley (*LCL*) 381–3.
164 *On the Cosmos* 397a30ff., trans. Furley (*LCL*) 383.
165 Trans. Furley (*LCL*) 395.
166 *On the Cosmos* 400a25ff., trans. Furley (*LCL*) 401.

167 Trans. Furley (*LCL*) 401. The authority of the poet is not only confined to the divine region; passages from the *Odyssey* (11.590, 11.589) are quoted (at 401a1ff.) as confirmation of the natural order divinely imposed on earth, with specific reference to various types of plants.

168 *On the Cosmos* 401a12ff., trans. Furley (*LCL*) 405–7.

169 See also West (1983) 218–19.

170 *On the Cosmos* 401b6–14, trans. Furley (*LCL*) 407–9.

171 Diogenes Laertius 7.151–4 provides an indication of the possible content of Stoic handbooks.

5 AN ENCYCLOPEDIC APPROACH

1 Seabrook (2000). See also Monmonier (2000). In England, 'the weather' established itself as a topic of acceptable polite conversation in (at least) the eighteenth century (Golinski (2000, 2001)), and is always felt to be a 'safe' topic.

2 Trans. Esolen (1995) 218.

3 For extracts and translation from Isidore, see Grant (1974) 25–7; on Isidore's use of Pliny, see Nauert (1980) 302.

4 See Fontaine (1995); Lettinck (1999); Schoonheim (2000).

5 See, e.g., B[urton] *The Surprizing Miracles of Nature and Art* (1683) facing 141 showing multiple suns (parhelia) observed during the Thirty Years War, 'The Terrible Prodigies during the Wars and Desolations in Germany'. See also the numerous illustrations reproduced in Schechner Genuth (1997).

6 Jenks (1983).

7 Janković (2000) 31–2.

8 Inwards (1893) 53 reproduces the ditty; another version features a shepherd rather than a sailor. See also Anderson (1994) 4.

9 Inwards (1893) ix.

10 Whether volcanic activity counted as meteorological in antiquity is not entirely clear, but it is worth noting that the poet who wrote about Aetna may have been Seneca's good friend Lucilius, to whom he addressed his meteorological *Natural Questions*, as well as many letters. See also Hine (2002).

11 Pliny the Younger *Letters* 6.16 (addressed to Tacitus the historian (*c.* 56–120 CE) in 104 CE), trans. B. Radice (*LCL* 1:427–9). See also *Letter* 6.20. See Hine (2002) 66 on Pliny the Younger's account of the volcanic eruption.

12 Degrassi (1963) 310; cf. Mingazzini (1928) 203. See also Lehoux (2000) 33–4.

13 How this fragment is to be understood has been debated; see Rehm (1949) col. 1302 and Lehoux (2000) 33–4.

14 Lehoux (2000) 60; cf. Degrassi (1963) 286f. On the *Menologia Rustica*, see Scullard (1981) 47; Broughton (1936). For an example of a fragment of a Latin calendar with precise stellar phases, see the Venusia Fasti, in Degrassi (1963) 55–62.

15 This sort of purely calendrical *parapēgma* is probably what was described by Cicero in two letters to Atticus. Cicero mentioned a παράπηγμα

ἐνιαύσιον in 5.14, which may correspond to the *clavus anni* in 5.15. See also Lehoux (2000) 67–9.

16 Lehoux (2000) 76 acknowledges the relation of Varro's text to the *parapēgma*, but does not think it is sufficiently alike to be called a *parapēgma*; he does include at 35–6 Columella's text in his list of extant astrometeorological *parapēgmata*.

17 Virgil *Georgics* 1.252–8, trans. Wilkinson (1982) 65.

18 Pliny *NH* 18.56.206, trans. Rackham (*LCL*) 5: 319. I use Rackham's (*LCL*) translation throughout this chapter.

19 The inclusion of Julius Caesar is noteworthy, since he was particularly interested in calendars and (apparently) employed the Greek astronomer Sosigenes to help devise the so-called Julian calendar, named after him. On Sosigenes, see Neugebauer (1975) 2: 575.

20 Trans. Rackham (*LCL*) 5: 317. Pliny *NH* 18.57.213 mentions an astronomical work by Hesiod; this may have had some relevance for meteorological topics. Fragments are preserved in quotations by other authors; see Hesiod *Fragmenta Selecta* (ed. Merkelbach and West) fragments 288–93.

21 See Bickerman (1980) 43–9 for an explanation of the naming and numbering of days in any given month.

22 Trans. Rackham (*LCL*) 5: 339–41.

23 Cf. Purcell (1996).

24 Pliny *NH* 18.57.210–17.

25 Pliny *NH* 18.57.212–13, trans. Rackham (*LCL*) 5: 325. These are the approximate dates for the ancient astronomers mentioned: Thales (sixth century BCE), Anaximander (sixth century BCE), Euctemon (fifth century BCE), Eudoxus (fourth century BCE).

26 In this context, it is also worth considering the first-century BCE Tower of the Winds in Athens; see Robinson (1943); Noble and Price (1968).

27 Pliny *NH* 2.45.117–18, trans. Rackham (*LCL*) 1: 259–61.

28 Pliny *NH* 2.45.116, trans. Rackham (*LCL*) 1: 257–9.

29 Pliny *NH* 2.44.114.

30 Pliny *NH* 2.46.119–2.48.129, trans. Rackham (*LCL*) 1: 261.

31 Pliny *NH* 2.48.130.

32 Pliny *NH* 2.49.131–2.50.134.

33 Pliny *NH* 2.57.147.

34 Pliny *NH* 2.58.148.

35 Pliny also notes, somewhat disparagingly, that 'the Greeks tell the story that Anaxagoras of Clazomenae in the 2nd year of the 78th Olympiad was enabled by his knowledge of astronomical literature to prophecy that in a certain number of days a rock would fall from the sun; and that this occurred in the daytime in the Goat's River district'. Pliny acknowledges that the brown stone, the size of a wagon-load, is still displayed. But, he remarks that 'if anyone believes in the fact of this prophecy, that involves his allowing that the divining powers of Anaxagoras covered a greater marvel, and that our understanding of the physical universe is annihilated and everything thrown into confusion if it is believed either that the sun is itself a stone or ever had a stone inside it'. While he is unwilling to believe that the sun contains stones, which may then fall to earth, Pliny

does not doubt that stones do fall frequently from the sky. He reports several such incidents of stones falling, and notes that he himself saw one in the territory of the Vocontii. (Pliny *NH* 2.59.149, trans. Rackham (*LCL*) 1: 285–7.)

36 Cf. to the Roman importation and worship of Magna Mater, described by Livy 29.11ff. Aristotle mentions (*Meteorology* 344b32ff.) the stone falling from the air at Aegospotami; cf. DK 59A1 (= Diogenes Laertius 2.6–15) and DK 59A11 (= Plutarch *Lysander* 12) on Anaxagoras.

37 Pliny *NH* 2.38.103–4, trans. Rackham (*LCL*) 1: 247.

38 Pliny *NH* 2.39.105–7, trans. Rackham (*LCL*) 1: 249.

39 Pliny *NH* 2.42.111, trans. Rackham (*LCL*) 1: 253.

40 Pliny *NH* 2.43.113, trans. Rackham (*LCL*) 1: 255.

41 Pliny *NH* 2.43.113.

42 Pliny *NH* 2.18.82.

43 Pliny *NH* 2.54.141, trans. Rackham (*LCL*) 1: 279.

44 Pliny *NH* 2.55.142–6.

45 Pliny *NH* 2.60.150–1.

46 Pliny *NH* 2.61.152, trans. Rackham (*LCL*) 1: 289.

47 Pliny *NH* 2.63.154, trans. Rackham (*LCL*) 1: 289.

48 Pliny *NH* 2.81.191–2, trans. Rackham (*LCL*) 1: 323–5.

49 Pliny *NH* 2.82.193–8.

50 Nauert (1980) 300; Nauert provides a description of manuscript transmission, printed editions and the commentary tradition. See also Gudger (1924); Kroll (1951) 430–3; Chibnall (1975).

51 Nauert (1980) 301.

52 See Kusukawa (2000) 125 citing Hermolao Barbaro's *Castigationes Plinianae* (Rome: E. Silber, 1493) and Niccolo Leoniceno's *On the Errors of Pliny and Others in Medicine* (Basle: H. Petri, 1529).

53 Kusukawa (2000) 125.

54 Nauert (1980) 316.

55 Janković (2000) 138.

56 Cf. Theophrastus, *On Winds* 47: 'The peculiar features of these winds can thus be rationally explained.' Trans. Wood (1894) 44.

57 Theophrastus *On Winds*, at 46, 49, 50; see also 51.

58 See Janković (2000) and Anderson (1994).

59 Robert Marc Friedman (1989) 164, citing V. Bjerknes to H. Bjerknes, 31 May 1918.

60 Customer reviews posted on Amazon.com for Robert B. Thomas (2002) *The Old Farmer's Almanac*, Dublin, NH: Yankee Publishing.

BIBLIOGRAPHY

Ancient authors

Note: The abbreviation *LCL* indicates a volume in the *Loeb Classical Library*.

Aelian *On the Characteristics of Animals*, trans. A.F. Scholfield, 3 vols, Cambridge, Mass.: Harvard University Press (*LCL*), 1958.

Aeschylus *Aeschylus,* trans. H.W. Smith, 2 vols, London: William Heinemann (*LCL*), 1926.

Aetna, trans. R. Ellis, Oxford: Clarendon Press, 1901.

—— in *Minor Latin Poets*, trans. J.W. Duff and A.M. Duff, 2 vols, London: William Heinemann (*LCL*), 1935.

Alexander of Aphrodisias *Alexandri in Meteorologicorum Libros Commentaria*, ed. M. Hayduck, *Commentaria in Aristotelem Graeca* III.2, Berlin: G. Reimer, 1899.

—— *On Aristotle Meteorology 4*, trans. E. Lewis, London: Duckworth, 1996.

Aratus *The Skies and Weather-Forecasts of Aratus*, trans. E. Poste, London: Macmillan, 1880.

—— *Phaenomena*, trans. G.R. Mair, in *Callimachus, Lycophron, Aratus*, Cambridge, Mass.: Harvard University Press (*LCL*), 1921.

—— *Phaenomena*, trans. Douglas Kidd, Cambridge: Cambridge University Press, 1997.

Aristophanes *The Acharnians The Clouds Lysistrata*, trans. Alan H. Sommerstein, London: Penguin, 1973.

Aristotle *The Complete Works of Aristotle, The Revised Oxford Translation*, ed. J. Barnes, 2 vols, Princeton: Princeton University Press, 1984, Bollingen Series 71.2.

—— *Metaphysics*, trans. W.D. Ross, in *The Complete Works of Aristotle*, vol. 2.

—— *Aristotelis Meteorologicorum Libri Quattuor*, ed. F.H. Fobes, Cambridge, Mass.: Harvard University Press, 1918.

—— *Meteorology*, trans. E.W. Webster, in *The Complete Works of Aristotle*, vol. 1.

—— *Meteorologica*, trans. H.D.P. Lee, Cambridge, Mass.: Harvard University Press (*LCL*), 1952.

—— *On Sleep*, trans. J.I. Beare, in *The Complete Works of Aristotle*, vol. 1.

——— *On the Soul*, trans. J.A. Smith, in *The Complete Works of Aristotle*, vol. 1.

——— *On Youth, Old Age, Life and Death, and Respiration*, trans. G.R.T. Ross, in *The Complete Works of Aristotle*, vol. 1.

——— *Parts of Animals*, trans. W. Ogle, in *The Complete Works of Aristotle*, vol. 1.

——— *Physics*, trans. R.P. Hardie and R.K. Gaye, in *The Complete Works of Aristotle*, vol. 1.

——— *Physics*, trans. P.H. Wicksteed and F.M. Cornford, 2 vols, Cambridge, Mass.: Harvard University Press (*LCL*), 1929–34.

——— *Politics*, trans. B. Jowett, in *The Complete Works of Aristotle*, vol. 2.

——— *Posterior Analytics*, trans. J. Barnes in *The Complete Works of Aristotle*, vol. 1.

——— *Prior Analytics*, trans. A.J. Jenkinson, in *The Complete Works of Aristotle*, vol. 1.

——— *Prior Analytics*, trans. R. Smith, Indianapolis: Hackett, 1989.

——— *Problems*, trans. E.S. Forster, in *The Complete Works of Aristotle*, vol. 2.

——— *Problems*, trans. W.S. Hett, 2 vols, London: Heinemann (*LCL*), 1936–37.

——— *Rhetoric*, trans. W. Rhys Roberts, in *The Complete Works of Aristotle*, vol. 2.

Avienus *Aviénus: Les Phénomènes d'Aratos*, ed. Jean Soubiran, Paris: Les Belles Lettres, 1981.

Cato *On Agriculture (de re rustica)*, trans. W.D. Hooper, rev. H.B. Ash, in *Marcus Porcius Cato: On Agriculture; Marcus Terentius Varro: On Agriculture*, Cambridge, Mass.: Harvard University Press (*LCL*), 1935.

Cicero *Letters to Atticus*, trans. E.O. Winstedt, 3 vols, London: William Heinemann (*LCL*), 1912–19.

——— *On Divination (de divinatione)*, trans. W.A. Falconer, in *Cicero, De Senectute, De Amicitia, De Divinatione*, Cambridge, Mass.: Harvard University Press (*LCL*), 1923.

——— *On the Nature of Gods (de natura deorum)*, trans. H. Rackham, in *Cicero, De Natura Deorum, Academica*, Cambridge, Mass.: Harvard University Press (*LCL*), 1933.

Columella *On Agriculture (de re rustica)*, trans. H.B. Ash (Books 1–4), E.S. Forster and E.H. Heffner (Books 5–12), in *Lucius Junius Moderatus Columella: On Agriculture*, 3 vols, Cambridge, Mass.: Harvard University Press (*LCL*), 1941–55.

Diogenes of Apollonius *Diogène d'Apollonie: La dernière cosmologie présocratique*, ed. A. Laks, Lille: Presses universitaires de Lille, 1983.

Diogenes Laertius *Lives of Eminent Philosophers*, trans. R.D. Hicks, 2 vols, Cambridge, Mass.: Harvard University Press (*LCL*), 1925.

Epicurus 'Letter to Pythocles', in *Diogenes Laertius: Lives of Eminent Philosophers* 10.3–11.7, trans. R.D. Hicks, 2 vols, Cambridge, Mass.: Harvard University Press (*LCL*), 1925, 2: 612–43.

Euripides *Euripides*, trans. A.S. Way, 4 vols, London: William Heinemann (*LCL*), 1912.

Geminus *Elementa astronomiae* (= *Isagoge*), ed. and German trans., C. Manitius, Stuttgart: Teubner, 1898.

—— *Gemini Elementorum Astronomiae*, ed. E.J. Dijksterhuis, Leiden: E.J. Brill, 1957.

—— *Géminos: Introduction aux Phénomènes*, ed. and French trans., G. Aujac, Paris: Les Belles Lettres, 1975.

'Geminus *parapēgma*' (= 'Calendarium') in Geminus *Elementa astronomiae* (= *Isagoge*), ed. C. Manitius, Stuttgart: Teubner, 1898, 210–33.

Germanicus *The Aratus Ascribed to Germanicus Caesar*, ed. and trans., D.B. Gain, London: The Athlone Press, 1976.

Herodotus *The Histories*, trans. A. de Sélincourt, London: Penguin, 1954, revised with Introduction and Notes by J. Marincola, 2003.

Hesiod *Fragmenta Selecta*, eds R. Merkelbach and M.L. West, Oxford: Clarendon Press, 1967.

—— *Hesiod: Homeric Hymns and Homerica*, trans. H.G. Evelyn-White, Cambridge, Mass.: Harvard University Press (*LCL*), 1914.

—— *Hesiod: The Poems and Fragments*, trans. A.W. Mair, Oxford: Clarendon Press, 1908.

—— *Hesiod: Works and Days*, ed. M.L. West, Oxford: Clarendon Press, 1978.

—— *Hesiod: The Works and Days, Theogony, The Shield of Herakles*, trans. R. Lattimore, Ann Arbor: University of Michigan Press, 1959.

—— *Hesiod: Works and Days, Theogony*, trans. S. Lombardo, Indianapolis: Hackett, 1993.

Hipparchus *Hipparchi in Arati et Eudoxi Phaenomena Commentariorum*, ed. C. Manitius, Leipzig: B.G. Teubner, 1894.

[Hippocrates] *Airs Waters Places*, in *Hippocrates*, vol. 1, trans. W.H.S. Jones, London: William Heinemann (*LCL*), 1923.

—— *Epidemics*, in *Hippocrates*, vol. 1, trans. W.H.S. Jones, London: William Heinemann (*LCL*), 1923.

—— *On Ancient Medicine*, in *Hippocrates*, vol. 1, trans. W.H.S. Jones, London: William Heinemann (*LCL*), 1923.

—— *Humours*, in *Hippocrates*, vol. 4, trans. W.H.S. Jones, London: William Heinemann (*LCL*), 1931.

Homer *Iliad*, trans. R. Lattimore, *The Iliad of Homer*, Chicago: University of Chicago Press, 1951, reprinted 1961.

—— *Odyssey*, trans. R. Lattimore, *The Odyssey of Homer*, New York: Harper & Row, 1965, reprinted 1975.

Livy *Livy*, vol. 8, trans. E.G. Moore, Cambridge, Mass.: Harvard University Press (*LCL*), 1949.

Lucretius *On the Nature of Things: De rerum natura*, ed. and trans. A.M. Esolen, Baltimore and London: Johns Hopkins University Press, 1995.

—— *On the Nature of Things (de rerum natura)*, trans. R. Melville, *Lucretius: On the Nature of the Universe*, Oxford: Clarendon Press, 1997.

—— *On the Nature of the Universe*, trans. R.E. Latham, Harmondsworth: Penguin, 1951.

John Lydus *Ioannis Lavrentii Lydi Liber de Ostentis et Calendaria Graeca Omnia*, ed. C. Wachsmuth, Leipzig: Teubner, 1897.

——— *Ioannes Lydus On Powers or the Magistracies of the Roman State*, ed. and trans. A.C. Bandy, Philadelphia: American Philosophical Society, 1983.

Manilius *Astronomica*, trans. Thomas Creech, *The Five Books of M. Manilius Containing a System of the Ancient Astronomy and Astrology: Together with the Philosophy of the Stoicks*, London, 1697.

——— *Astronomica*, trans. G.P. Goold, Cambridge, Mass.: Harvard University Press (*LCL*), 1977.

Olympiodorus *Olympiodori in Aristotelis Meteora Commentaria*, ed. G. Stüve, *Commentaria in Aristotelem Graeca* XII.2, Berlin: G. Reimer, 1900.

John Philoponus *Ioannis Philoponi in Aristotelis Meteorologicorum Librum Primum Commentarium*, ed. M. Hayduck, *Commentaria in Aristotelem Graeca* XIV.1, Berlin: G. Reimer, 1901.

Plato *Timaeus*, trans. B. Jowett, in *Plato: The Collected Dialogues*, eds E. Hamilton and H. Cairns, Princeton: Princeton University Press, Bollingen Series 71, 1961; reprinted from *The Dialogues of Plato*, 4th edn, 1953; 1st edn 1871.

——— *Plato: Complete Works*, eds J.M. Cooper, D.S. Hutchinson and J. Barnes, Indianapolis: Hackett, 1997.

Pliny the Elder *Natural History*, trans. H. Rackham *et al.*, *Pliny: Natural History*, 10 vols, Cambridge, Mass.: Harvard University Press (*LCL*), 1938–63.

Pliny the Younger *Letters of the Younger Pliny*, trans. B. Radice, London: William Heinemann (*LCL*), 1969.

Plutarch *Plutarch's Lives*, trans. B. Perrin, 11 vols, London: William Heinemann (*LCL*), 1914–26, vol. 4.

[Pseudo-Aristotle] *On the Cosmos* (= *De mundo*), trans. D.J. Furley, in *Aristotle: On Sophistical Refutations, On Coming-to-be and Passing-away, On the Cosmos*, Cambridge, Mass.: Harvard University Press (*LCL*), 1955.

——— *The Situations and Names of Winds*, trans. E.S. Forster, in *The Complete Works of Aristotle*, vol. 2.

[Pseudo-Plato] *The Axiochus: On Death and Immortality, a Platonic Dialogue*, trans. E.H. Blakeney, London: Frederick Muller, 1937.

——— *Axiochus*, trans. J.P. Hershbell, Chico, CA: Scholars Press, Society of Biblical Literature, Texts and Translations 21, Graeco-Roman Religion Series 6, 1981.

Ptolemy *Harmonics*, trans. A. Barker, in *Greek Musical Writings II*, Cambridge: Cambridge University Press, 1989.

——— *Phases*, in *Opera astronomica minora*, ed. J.L. Heiberg, Leipzig: Teubner, 1907, 1–67.

——— *Syntaxis mathematica*, ed. J.L. Heiberg, 2 vols, Leipzig: Teubner, 1898–1903.

——— *Ptolemy's Almagest*, trans. G. Toomer, New York: Springer Verlag, 1984.

——— *Tetrabiblos*, eds F. Boll and A. Boer, Leipzig: Teubner, 1957.

—— *Tetrabiblos*, trans. F.E. Robbins, Cambridge, Mass.: Harvard University Press (*LCL*), 1980.

Seneca *Letters to Lucilius* (*Ad Lucilium epistulae morales*), trans. R.M. Gummere, in *Seneca, Epistles*, 3 vols, Cambridge, Mass.: Harvard University Press (*LCL*), 1917–25.

—— *Natural Questions*, trans. T.H. Corcoran, *Seneca, Naturales Quaestiones*, 2 vols, London: William Heinemann (*LCL*), 1971–72.

—— *To Helvia on Consolation* in *Seneca: Moral Essays*, 3 vols, trans. J.W. Basore, Cambridge, Mass.: Harvard University Press (*LCL*), 1928–35, vol. 2.

Sextus Empiricus *Against the Professors* (= *Adversus Mathematicos*), trans. R.G. Bury, *Sextus Empiricus*, 4 vols, London: William Heinemann (*LCL*), 1933–49, vol. 3.

Theophrastus *Meteorology/Metarsiology*, in H. Daiber (1992) 'The *Meteorology* of Theophrastus in Syriac and Arabic Translation', in Fortenbaugh, W.W. and Gutas, D. (eds) *Theophrastus: His Psychological, Doxographical, and Scientific Writings*, New Brunswick, NJ: Transaction Publishers, *Rutgers University Studies in Classical Humanities*, vol. 5, 166–293.

—— *Metaphysics*, trans. W.D. Ross and F.H. Fobes, Oxford: Clarendon Press, 1929.

—— *Metaphysics*, trans. M. van Raalte, Leiden: E.J. Brill, 1993.

—— *Théophraste: Métaphysique*, eds A. Laks and G.W. Most, Paris: Les Belles Lettres, 1993.

[Theophrastus] *Concerning Weather Signs*, in *Theophrastus: Enquiry into Plants and Minor Works, On Odours and Weather Signs*, 2 vols, trans. A. Hort, Cambridge, Mass.: Harvard University Press (*LCL*), 1916–26; reprinted 1961, vol. 2, 390–433.

—— *On Weather Signs*, trans. J.G. Wood, in *Theophrastus of Eresus: On Winds and On Weather Signs*, London: Edward Stanford, 1894.

—— *On Signs*, eds D. Sider and W. Brunschön (forthcoming).

Varro *On Agriculture* (*de re rustica*), trans. W.D. Hooper, rev. H.B. Ash, in *Marcus Porcius Cato: On Agriculture; Marcus Terentius Varro: On Agriculture*, Cambridge, Mass.: Harvard University Press (*LCL*), 1935.

Virgil *Georgics*, trans. C. Day Lewis, in *The Eclogues, The Georgics*, trans. 1940; Oxford: Oxford University Press; reprinted 1999.

—— *Georgics*, trans. H.R. Fairclough, in *Virgil: Eclogues, Georgics, Aeneid I–VI*, Cambridge, Mass.: Harvard University Press (*LCL*), 1916.

—— *The Georgics*, trans. L.P. Wilkinson, Harmondsworth: Penguin, 1982.

Vitruvius *On Architecture* (*de architectura*), trans. F. Granger, 2 vols, Cambridge, Mass.: Harvard University Press (*LCL*), 1931–4.

Modern authors

Algra, K., Barnes, J., Mansfeld, J. and Schofield, M. (eds) (1999) *The Cambridge History of Hellenistic Philosophy*, Cambridge: Cambridge University Press.

Algra, K.A., Koenen, M.H. and Schrijvers, P.H. (1997) *Lucretius and his Intellectual Background*, Amsterdam: Koninklijke Nederlandse Akademie van Wetenschappen Verhandelingen, Afd. Letterkunde, Nieuwe reeks, deel 172.

Allan, D.J. (1936) Review of H. Strohm (1935) *Untersuchungen zur Entwicklungsgeschichte der aristotelischen Meteorologie*, *Philologus*, Supplementband 28, Leipzig, in *Classical Review* 50: 37.

—— (1961) 'Quasi-mathematical method in the *Eudemian Ethics*', in Mansion, S. (ed.) *Aristote et les problèmes de méthode*, Louvain: Publications Universitaires de Louvain, 303–18.

Allen, J. (2001) *Inference from Signs: Ancient Debates about the Nature of Evidence*, Oxford: Clarendon Press.

Amsler, M. (1989) *Etymology and Grammatical Discourse in Late Antiquity and the Early Middle Ages*, Amsterdam: John Benjamins, *Amsterdam Studies in the Theory and History of Linguistic Science* 44.

Anderson, K.M. (1994) 'Practical science: meteorology and the forecasting controversy in mid-Victorian Britain', unpublished doctoral thesis, Northwestern University.

Annas, J. (1987) 'Die Gegenstände der Mathematik bei Aristoteles', in Graeser, A. (ed.) *Mathematics and Metaphysics in Aristotle/Mathematik und Metaphysik bei Aristoteles: Akten des X. Symposium Aristotelicum Sigriswil, 6–12 September 1984*, Bern: Paul Haupt, 131–47.

Asmis, E. (1984) *Epicurus' Scientific Method*, Ithaca: Cornell University Press.

—— (1996) 'Epicurean semiotics', in Manetti, G. (ed.) *Knowledge Through Signs: Ancient Semiotic Theories and Practices*, Turhout: Brepols, 155–86.

—— (1999) 'Epicurean epistemology', in Algra, K., Barnes, J., Mansfeld, J. and Schofield, M. (eds) *The Cambridge History of Hellenistic Philosophy*, Cambridge: Cambridge University Press, 260–94.

Aujac, G. (1975) *Géminos: Introduction aux Phénomènes*, Paris: Les Belles Lettres.

Balme, D.M. (1972) *Aristotle's* De Partibus Animalium *I and* De Generatione Animalium *I*, Oxford: Clarendon Press.

Bandy, A.C. (1983) (ed. and trans.) *Ioannes Lydus On Powers or the Magistracies of the Roman State*, Philadelphia: American Philosophical Society.

Barker, A. (2000) *Scientific Method in Ptolemy's 'Harmonics'*, Cambridge: Cambridge University Press.

Barnes, J. (1975a) *Aristotle's Posterior Analytics*, Oxford: Clarendon Press.

—— (1975b) 'Aristotle's theory of demonstration', in Barnes, J., Schofield, M. and Sorabji, R. (eds) *Articles on Aristotle*, 1, *Science*, London: Duckworth, 65–87.

—— (1977) Review of G. Reale (1974) *Aristotele: Trattato sul Cosmo per Alessandro*, Naples: Loffredo, in *Classical Review* (New Series) 27: 40–3.

—— (1980) 'Aristotle and the methods of ethics', *Revue Internationale de Philosophie* 133–4: 490–511.

—— (ed.) (1995a) *The Cambridge Companion to Aristotle*, Cambridge: Cambridge University Press.

——— (1995b) 'Metaphysics', in Barnes, J. (ed.) *The Cambridge Companion to Aristotle*, Cambridge: Cambridge University Press, 66–108.

Barnes, J., Brunschwig, J., Burnyeat, M. and Schofield, M. (eds) (1982) *Science and Speculation: Studies in Hellenistic Theory and Practice*, Cambridge: Cambridge University Press.

Barton, T. (1994a) *Ancient Astrology*, London: Routledge.

——— (1994b) *Power and Knowledge: Astrology, Physiognomics and Medicine under the Roman Empire*, Ann Arbor: University of Michigan Press.

Beagon, M. (1989) *Roman Nature: The Thought of Pliny the Elder*, Oxford: Clarendon Press.

Beard, M. (1986) 'Cicero and divination: the formation of a Latin discourse', *Journal of Roman Studies* 76: 33–46.

Beazley, J.D. (1963) *Attic Red-figure Vase-painters*, 2nd edn, Oxford: Clarendon Press.

Bekkum, W. van, Houben, J., Sluiter, I. and Versteegh, K. (1997) *The Emergence of Semantics in Four Linguistic Traditions: Hebrew, Sanskrit, Greek, Arabic*, Amsterdam: John Benjamins, *Amsterdam Studies in the Theory and History of Linguistic Science* 82.

Bickerman, E.J. (1980) *Chronology of the Ancient World*, revised edn, London: Thames and Hudson.

Bing, P. (1993) 'Aratus and his audiences', in Schiesaro, A., Mitsis, P. and Clay, J.S. (eds) *Mega Nepios: il destinatario nell'epos didascalico*, Pisa: Giardini, *Materiali e discussioni per l'analisi dei testi classici* 31, 99–109.

Blänsdorf, J. (ed.) (1995) *Fragmenta Poetarum Latinorum epicorum et lyricorum praeter Ennium et Lucilium*, 3rd edn, Stuttgart: Teubner.

Boardman, J. (1975) *Athenian Red Figure Vases: The Archaic Period*, London: Thames and Hudson.

Böckh, A. (1863) *Über die vierjährigen Sonnenkreise der Alten, vorzüglich den Eudoxischen. Ein Beitrag zur Geschichte der Zeitrechnung und des Kalenderwesens der Aegypter, Griechen und Römer*, Berlin: Georg Reimer.

Boer, J.Z. de, Hale, J.R. and Chanton, J. (2001) 'New evidence for the geological origins of the ancient Delphic oracle (Greece)', *Geology* 29 no. 8 (August): 707–10.

Boeuffle, A. Le (1973) 'Les Aratea de Germanicus', *Revue de Philologie* 47: 61–7.

Böker, R. (1958) 'Winde', *Paulys Realencyclopädie der classischen Altertumswissenschaft* 8A2, columns 2215–65, Abteilung B: Die geophysischen Windtheorien im Altertum.

——— (1962a) 'Wetterzeichen', *Paulys Realencyclopädie der classischen Altertumswissenschaft*, Supplementband 9, columns 1609–92.

——— (1962b) 'Windfristen'. *Paulys Realencyclopädie der classischen Altertumswissenschaft*, Supplementband 9, columns 1697–705.

Bos, G. and Burnett, C. (2000) *Scientific Weather Forecasting in the Middle Ages: The Writings of Al-Kindī*, London: Kegan Paul International.

Bourgey, L. (1975) 'Observation and experiment in analogical explanation', in Barnes, J., Schofield, M. and Sorabji, R. (eds) *Articles on Aristotle 1. Science*. London: Duckworth, 175–82.

Bowen, A.C. and Goldstein, B.R. (1988) 'Meton of Athens and astronomy in the late fifth century B.C.', in Leichty, E., Ellis, M. de J. and Gerardi, P. (eds) *A Scientific Humanist: Studies in Memory of Abraham Sachs*, Philadelphia: Occasional Publications of the Samuel Noah Kramer Fund, 9, 39–82.

Bowman, A.K. (1991) 'Literacy in the Roman empire: mass and mode', in *Literacy in the Ancient World, Journal of Roman Archaeology, Supplementary Series* 3: 119–31.

Boyer, Carl B. (1959) *The Rainbow: From Myth to Mathematics*, Princeton: Princeton University Press, 1987 reprint.

Brok, M.F.A. (1995) *L. Annaeus Seneca: Naturwissenschaftliche Untersuchungen*, Darmstadt: Wissenschaftliche Buchgesellschaft.

Broughton, A.L. (1936) 'The *Menologia Rustica*', *Classical Philology* 31: 354–6.

Brown, D. (2000) *Mesopotamian Planetary Astronomy-astrology*, Groningen: Styx, Cuneiform Monographs 18.

Brückner, A. (1931) 'Mitteilungen aus dem Kerameikos V. Vorbericht über Ergebnisse der Grabung 1929', *Mitteilungen des deutschen Archäologischen Instituts*, Athenische Abteilung 56: 1–32.

Brunt, P.A. (1980) 'On historical fragments and epitomes', *Classical Quarterly* (New Series) 30: 477–94.

Burnett, C. (1993) 'An unknown Latin version of an ancient *parapēgma*: the weather-forecasting stars in the *Iudicia* of pseudo-Ptolemy', in Anderson, R.G.W., Bennett, J.A. and Ryan, W.F. (eds) *Making Instruments Count: Essays on Historical Scientific Instruments presented to Gerard L'Estrange Turner*, Aldershot: Variorum.

Burnyeat, M. (1981) 'Aristotle on knowledge', in Berti, E. (ed.) *Aristotle on Science: The Posterior Analytics*, Proceedings of the Eighth Symposium Aristotelicum, Padua: Editrice Antenore, *Studia Aristotelica* 9, 97–139.

—— (1982) 'The origins of non-deductive inference', in Barnes, J., Brunschwig, J., Burnyeat, M. and Schofield, M. (eds) *Science and Speculation: Studies in Hellenistic Theory and Practice*, Cambridge: Cambridge University Press, 193–238.

—— (1994) 'Enthymeme: Aristotle on the logic of persuasion', in Furley, D.J. and Nehamas, A. (eds) *Aristotle's Rhetoric, Proceedings of the Twelfth Symposium Aristotelicum*, Princeton: Princeton University Press, 3–55.

Burnyeat, M. and Schofield, M. (eds) (1982) *Science and Speculation: Studies in Hellenistic Theory and Practice*, Cambridge: Cambridge University Press.

B[urton], R. [pseudonym] (1683) *The Surprizing Miracles of Nature and Art*, London: Nath. Crouch (compiler).

Cameron, A. (1967) 'Macrobius, Avienus, and Avianus', *Classical Quarterly* 17: 385–99.

—— (1995) 'Avienus or Avienius?', *Zeitschrift für Papyrologie und Epigraphik* 108: 252–62.

Capelle, W. (1905) 'Die Schrift von der Welt', *Neue Jahrbücher für das klassische Altertum* 15: 529–68.

—— (1912) 'Das Proömium der Meteorologie', *Hermes* 47: 514–35.

—— (1913) 'Zur Geschichte der meteorologischen Literatur', *Hermes* 48: 321–58.

—— (1914) 'Die Nilschwelle', *Neue Jahrbücher für das klassische Altertum, Geschichte und deutsche Literatur und für Pädagogik* 33: 317–61.

—— (1916) 'Berges- und Wolkenhöhen bei griechischen Physikern', *ΣΤΟΙΧΕΙΑ (Stoicheia)* 5: 1–47.

—— (1935) 'Meteorologie', *Paulys Real-Encyclopädie der classischen Altertums-wissenschaft*, Supplementband 6: 315–58.

Cartledge, P.A. and Sallares, J.R. (1996) 'Earthquakes', in Hornblower, S. and Spawforth, A. (eds) *The Oxford Classical Dictionary*, 3rd edn, Oxford: Oxford University Press.

Cherniss, H. (1936) Review of H. Strohm (1935) *Untersuchungen zur Entwicklungsgeschichte der aristotelischen Meteorologie, Philologus*, Supplement-band 28, Leipzig, in *American Journal of Philology* 57: 371–4.

Chibnall, M. (1975) 'Pliny's *Natural History* and the Middle Ages', in Dorey, T.A. (ed.) *Empire and Aftermath: Silver Latin II*, London: Routledge & Kegan Paul, 57–78.

Ciarallo, A. and Carolis, E. De (1999) *Pompeii: Life in a Roman Town*, Milan: Electa.

Clarke, K. (1999) *Between Geography and History: Hellenistic Constructions of the Roman World*, Oxford: Clarendon Press.

Clay, J.S. (1993) 'The education of Perses: from "Mega Nepios" to "Dion Genos" and back', in Schiesaro, A., Mitsis, P. and Clay, J.S. (eds) *Mega Nepios: il destinatario nell'epos didascalico*, Pisa: Giardini, *Materiali e discussioni per l'analisi dei testi classici* 31, 23–33.

Cleary, J.J. (1995) *Aristotle and Mathematics: Aporetic Method in Cosmology and Metaphysics*, Leiden: E.J. Brill.

Codoñer, C. (1989) 'La physique de Sénèque: Ordonnance et structure des "Naturales Quaestiones"', *Aufstieg und Niedergang der römischen Welt (ANRW)* 2.36.3: 1789–822.

The Compact Edition of the Oxford English Dictionary (1971) 2 vols, Oxford: Oxford University Press.

Connors, C. (1997) 'Field and forum: culture and agriculture in Roman rhetoric', in Dominick, W.J. (ed.) *Roman Eloquence: Rhetoric in Society and Literature*, London: Routledge.

Conte, G.B. (1991) *Generi e lettori: Lucrezio, L'elegia d'amore, L'enciclopedia di Plinio*, trans. G.W. Most (1994) *Genres and Readers: Lucretius, Love Elegy, Pliny's Encyclopedia*, Baltimore: Johns Hopkins University Press.

Cook, A.B. (1914–40) *Zeus: A Study in Ancient Religion*, 3 vols, Cambridge: Cambridge University Press.

Cook, K.C. (1989) 'The underlying thing, the underlying nature and matter: Aristotle's analogy in physics I 7', in Penner, T. and Kraut, R. (eds) *Nature, Knowledge and Virtue: Essays in Memory of Joan Kung, Apeiron* 23: 105–19.

Cooper, J.M. (1982) 'Aristotle on natural theology', in Schofield, M. and Nussbaum, M.C. (eds) *Language and Logos: Studies in Ancient Greek Philosophy presented to G.E.L. Owen*, Cambridge: Cambridge University Press, 197–222.
—— (1985) 'Hypothetical necessity', in Gotthelf, A. (ed.) *Aristotle on Nature and Living Things: Philosophical and Historical Studies presented to David M. Balme on his Seventieth Birthday*, Pittsburgh: Mathesis, 151–67.
Corcoran, T.H. (1971/72) *Seneca, Naturales Quaestiones*, 2 vols, London: William Heinemann (*LCL*), vol. 1 (1971), vol. 2 (1972).
Courtney, E. (1993) (ed.) *The Fragmentary Latin Poets*, Oxford: Clarendon Press.
—— (1996) 'Terentius Varro Atacinus, Publius', in Hornblower, S. and Spawforth, A. (eds) *The Oxford Classical Dictionary*, 3rd edn, Oxford: Oxford University Press.
Coutant, V. and Eichenlaub, V.L. (1975) *Theophrastus: De Ventis*, Notre Dame: University of Notre Dame Press.
[Creech, Thomas] (1697) *The Five Books of M. Manilius, Containing a System of Ancient Astronomy and Astrology: Together with the Philosophy of the Stoicks*, London: Jacob Tonson.
Cronin, P. (1992) 'The authorship and sources of the *Peri semeion* ascribed to Theophrastus', in Fortenbaugh, W.W. and Gutas, D. (eds) *Theophrastus: His Psychological, Doxographical, and Scientific Writings*, New Brunswick, NJ: Transaction Publishers, *Rutgers University Studies in Classical Humanities*, vol. 5, 307–45.
—— (2001) 'Weather lore as a source of Homeric imagery', *ΕΛΛΗΝΙΚΑ (Hellenika)* 51: 7–24.
Daiber, H. (1992) 'The *Meteorology* of Theophrastus in Syriac and Arabic translation', in Fortenbaugh, W.W. and Gutas, D. (eds) *Theophrastus: His Psychological, Doxographical, and Scientific Writings*, New Brunswick, NJ: Transaction Publishers, *Rutgers University Studies in Classical Humanities*, vol. 5, 166–293.
Dalzell, A. (1996) *The Criticism of Didactic Poetry: Essays on Lucretius, Virgil, and Ovid*, Toronto: University of Toronto Press.
Degrassi, A. (1963) *Inscriptiones Italiae* 13, fasc. 2, Rome: Libreria dello Stato.
Denyer, N. (1985) 'The case against divination: an examination of Cicero's *De Divinatione*', *Proceedings of the Cambridge Philological Society* 211 (New Series 31): 1–10.
Dessau, H. (1904) 'Zu den Milesischen Kalenderfragmenten', *Sitzungsberichte der königlich preussischen Akademie der Wissenschaften, philosophisch-historischen Classe* 23: 266–8.
Dicks, D.R. (1970) *Early Greek Astronomy to Aristotle*, Ithaca, NY: Cornell University Press.
Diels, H. (1879) *Doxographi Graeci*, Berlin: G. Reimer.
Diels, H. and Kranz, W. (eds) (1952) *Fragmente der Vorsokratiker*, 6th edn, 3 vols, Berlin: Weidmann.
Diels, H. and Rehm, A. (1904) 'Parapegmenfragmente aus Milet', *Sitzungsberichte der königlich preussischen Akademie der Wissenschaften, philosophisch-historischen Classe* 23: 92–111.

Dilke, O.A.W. (1987) 'Itineraries and geographical maps in the early and late Roman empires', in Harley, J.B. and Woodward, D. (eds) *The History of Cartography*, vol. I: *Cartography in Prehistoric, Ancient, and Medieval Europe and the Mediterranean*, Chicago: University of Chicago Press, 234–57.

Drossaart-Lulofs, H.J (1955) 'The Syriac translation of Theophrastus' Meteorology', in *Autour d' Aristote. Recueil d'études de philosophie ancienne et médiévale offert à Monseigneur A. Mansion*, Louvain: Publications universitaires de Louvain, Bibliothèque philosophique de Louvain 16, 433–49.

Duff, J.W. and Duff, A.M. (1935) *Minor Latin Poets*, 2 vols, London: William Heinemann (*LCL*).

Edelstein, L. and Kidd, I.G. (1972) *Posidonius I. The Fragments*, Cambridge: Cambridge University Press.

Effe, B. (1977) *Dichtung und Lehre: Untersuchungen zur Typologie des antiken Lehrgedichts*, Munich: C.H. Beck'sche Verlagsbuchhandlung, *Zetemata* 69.

Eichholz, D.E. (1949) 'Aristotle's theory of metals and minerals', *Classical Quarterly* 43: 141–6; reprinted as 38–47 in Eichholz (1965).

—— (1965) *Theophrastus: De Lapidibus*, Oxford: Clarendon Press.

Evans, J. (1983) 'The history and practice of ancient astronomy', unpublished doctoral thesis, University of Washington.

—— (1998) *The History and Practice of Ancient Astronomy*, Oxford: Oxford University Press.

—— (1999) 'The material culture of Greek astronomy', *Journal for the History of Astronomy* 30: 237–307.

Evelyn-White, H.G. (1914) *Hesiod: The Homeric Hymns and Homerica*, Cambridge, Mass.: Harvard University Press (*LCL*).

Fairclough, H.R. (1916) *Virgil: Georgics*, in *Virgil: Eclogues, Georgics, Aeneid I–VI*, Cambridge, Mass.: Harvard University Press (*LCL*).

Falconer, W.A. (1923) *Cicero, De Divinatione*, in *Cicero, De Senectute, De Amicitia, De Divinatione*, Cambridge, Mass.: Harvard University Press (*LCL*).

Farrell, J. (1991) *Vergil's Georgics and the Traditions of Ancient Epic: The Art of Allusion in Literary History*, New York: Oxford University Press.

Ferguson, J. (1972) 'Seneca the man', in Dudley, D.R. (ed.) *Neronians and Flavians: Silver Latin I*, London: Routledge & Kegan Paul, 1–61.

Fiedler, W. (1978) *Analogiemodelle bei Aristoteles: Untersuchungen zu den Vergleichen zwischen den einzelnen Wissenschaften und Künsten*, Amsterdam: B.R. Grüner.

Fine, G. (1987) 'Forms as causes: Plato and Aristotle', in Graeser, A. (ed.) *Mathematics and Metaphysics in Aristotle/Mathematik und Metaphysik bei Aristoteles: Akten des X. Symposium Aristotelicum Sigriswil, 6–12 September 1984*, Bern: Paul Haupt.

Fisher, R. (1977) 'Astronomy and the calendar in Hesiod', *Echos du Monde Classique, Classical News and Views* 21: 58–63.

Flashar, H. (1962) *Aristoteles: Problemata Physica (Aristoteles Werke in deutscher Übersetzung*, vol. 19), Darmstadt: Wissenschaftliche Buchgesellschaft.

Fobes, F.H. (1916) 'A note on Aristotle *Meteorology* II.6', *Classical Review* 30: 48–9.

Fontaine, R. (1995) *Otot ha-shamayim: Samuel Ibn Tibbon's Hebrew version of Aristotle's Meteorology*, Leiden: E.J. Brill.

—— (2000) 'Between scorching heat and freezing cold: medieval Jewish authors on the inhabited and uninhabited parts of the earth', *Arabic Sciences and Philosophy* 10: 101–37.

—— (2001) 'The reception of Aristotle's *Meteorology* in Hebrew scientific writings of the thirteenth century', *Aleph* 1: 101–39.

Fortenbaugh, W.W. and Gutas, D. (1992) *Theophrastus: His Psychological, Doxographical and Scientific Writings*, New Brunswick: Transaction, *Rutgers University Studies in Classical Humanities*, vol. 5.

Fortenbaugh, W.W., Huby, P.M. and Long, A.A. (1985) *Theophrastus of Eresus: On His Life and Work*, New Brunswick: Transaction, *Rutgers University Studies in Classical Humanities*, vol. 2.

Fortenbaugh, W.W., Huby, P.M., Sharples, R.W. and Gutas, D. (1992) *Theophrastus of Eresus: Sources for his Life, Writings, Thought and Influence*, 2 vols, Leiden: Brill, *Philosophia Antiqua* 54.

Fowler, D. (1999) *The Mathematics of Plato's Academy: A New Reconstruction*, 2nd edn, Oxford: Clarendon Press.

Fowler, D.H. and Turner, E.G. (1983) 'Hibeh Papyrus i 27: an early example of Greek arithmetical notation', *Historia Mathematica* 10: 344–59.

Fowler, P.G. and Fowler, D.P. (1996a) 'Lucretius (Titus Lucretius Carus)', in Hornblower, S. and Spawforth, A. (eds) *The Oxford Classical Dictionary*, 3rd edn, Oxford: Oxford University Press.

—— (1996b) 'Virgil (Publius Vergilius Maro)' in Hornblower, S. and Spawforth, A. (eds) *The Oxford Classical Dictionary*, 3rd edn, Oxford: Oxford University Press.

—— (1997) 'Introduction' to *Lucretius: On the Nature of the Universe*, trans. Ronald Melville, Oxford: Clarendon Press.

Fraassen, B.C. van (1980a) 'A re-examination of Aristotle's philosophy of science', *Dialogue* 19: 20–45.

—— (1980b) *The Scientific Image*, Oxford: Clarendon Press.

Fränkel, H. (1975) *Early Greek Poetry and Philosophy: A History of Greek Epic, Lyric, and Prose to the Middle of the Fifth Century*, trans. M. Hadas and J. Willis, Oxford: Basil Blackwell; trans. of *Dichtung und Philosophie des frühen Griechentums*, Munich: C.H. Beck'sche Verlagsbuchhandlung, 1962, a revision of the 1951 edn.

Franklin, J.L. Jr (1991) 'Literacy and the parietal inscriptions of Pompeii', in *Literacy in the Ancient World, Journal of Roman Archaeology, Supplementary Series* 3: 77–98.

Freeland, C.A. (1990) 'Scientific explanation and empirical data in Aristotle's *Meteorology*', *Oxford Studies in Ancient Philosophy* 8: 67–102.

—— (1991) 'Accidental causes and real explanations', in Judson, L. (ed.) *Aristotle's* Physics: *A Collection of Essays*, Oxford: Clarendon Press, 49–72.

French, R. (1994) *Ancient Natural History: Histories of Nature*, London: Routledge.

French, R. and Greenaway, F. (eds) (1986) *Science in the Early Roman Empire: Pliny the Elder, his Sources and Influence*, London: Croom Helm.

Freudenthal, G. (1995) *Aristotle's Theory of Material Substance: Heat and Pneuma, Form and Soul*, Oxford: Clarendon Press.

Friedman, R.M. (1989) *Appropriating the Weather: Vilhelm Bjerknes and the Construction of a Modern Meteorology*, Ithaca, NY: Cornell University Press; reprinted 1993.

Frisinger, H. (1977) *The History of Meteorology: to 1800*, New York: Science History Publications.

Furley, D.J. (1955) *Aristotle: On the Cosmos*, in *Aristotle: On Sophistical Refutations, On Coming-to-be and Passing-away, On the Cosmos*, Cambridge, Mass.: Harvard University Press (*LCL*).

—— (1983) 'The mechanics of the Meteorologica IV. A prolegomenon to biology', in Moraux, P. and Wiesner, J. (eds) *Zweifelhaftes im Corpus Aristotelicum: Studien zu einigen Dubia. Akten des 9. Symposium Aristotelicum (Berlin 7–16 September 1981)*, Berlin: Walter de Gruyter.

—— (1985) 'The rainfall example in *Physics* ii 8', in Gotthelf, A. (ed.) *Aristotle on Nature and Living Things: Philosophical and Historical Studies presented to David M. Balme on his Seventieth Birthday*, Pittsburgh: Mathesis Publications, 177–82.

Gain, D.B. (1976) *The Aratus Ascribed to Germanicus Caesar*, London: The Athlone Press.

Gale, M.R. (2000) *Virgil on the Nature of Things: The* Georgics, *Lucretius and the Didactic Tradition*, Cambridge: Cambridge University Press.

Gee, E. (2000) *Ovid, Aratus and Augustus: Astronomy in Ovid's* Fasti, Cambridge: Cambridge University Press.

Gilbert, O. (1907) *Die meteorologischen Theorien des griechischen Altertums*, Leipzig: Teubner; reprinted 1967, Hildesheim: Georg Olms.

Gillespie, W.E. (1938) *Vergil, Aratus and Others: The Weather-Sign as a Literary Subject*, Princeton: printed by Edwards Brothers, Ann Arbor, Michigan.

Gjerstad, E. (1953) 'Lunar months of Hesiod and Homer', *Opuscula Atheniensia, Skrifter Utgivna av Svenska Institutet i Athen*, 4, II, Lund: CWK Gleerup.

Goldstein, B. and Bowen, A. (1983) 'A new view of early Greek astronomy', *Isis* 74: 330–40.

Golinski, J. (2000) ' "Weather, fashions, news and the like publick topics": meteorology and modernity in eighteenth-century Britain', paper presented at a Seminar at the Max Planck Institute for History of Science, Berlin, November 2000.

—— (2001) 'Climates of enlightenment: time, talk, and the weather in eighteenth-century Britain', paper presented at a Seminar in the Department of History and Philosophy of Science, University College, London, February 2001, and at a Seminar in the Department of History and Philosophy of Science, Cambridge University, May 2001.

Goody, J. (1977) *The Domestication of the Savage Mind*, Cambridge: Cambridge University Press.

Goodyear, F.R.D. (1984) 'The "Aetna": thought, antecedents, and style', in *Aufstieg und Niedergang der römischen Welt (ANRW)* 2.32.1: 344–63.

Goold, G.P. (1977) *Manilius: Astronomica*, Cambridge, Mass.: Harvard University Press (*LCL*).

Gottschalk, H.B. (1961) 'The authorship of *Meteorologica*, Book IV', *Classical Quarterly* (New Series) 11: 67–79.

—— (1987) 'Aristotelian philosophy in the Roman world', in *Aufstieg und Niedergang der römischen Welt (ANRW)* 2.36.2: 1079–174.

Grant, E. (1974) *A Source Book in Medieval Science*, Cambridge, Mass.: Harvard University Press.

Graßhoff, G. (1993) 'The Babylonian tradition of celestial phenomena and Ptolemy's fixed star calendar', in Galter, H.D. (ed.) *Die Rolle der Astronomie in den Kulturen Mesopotamiens*, *Grazer Morgenländische Studien* 3: 95–134.

—— (1999) 'Normal star observations in late Babylonian astronomical diaries', in Swerdlow, N.M. (ed.) *Ancient Astronomy and Celestial Divination*, Cambridge, Mass.: MIT Press, 97–147.

Graver, M. (1999) Commentary on Inwood, B. (1999) 'God and human knowledge in Seneca's Natural Questions', *Proceedings of the Boston Area Colloquium in Ancient Philosophy* 15: 44–54.

Greenler, R. (1980) *Rainbows, Halos and Glories*, Cambridge: Cambridge University Press.

Grene, M. (1985) 'About the division of the sciences', in Gotthelf, A. (ed.) *Aristotle on Nature and Living Things: Philosophical and Historical Studies presented to David M. Balme on his Seventieth Birthday*, Pittsburgh: Mathesis, 9–13.

Grenfell, B.P. and Hunt, A.S. (1906) *The Hibeh Papyri. Part I*, London: Egypt Exploration Fund.

Griffin, M.T. (1976/92) *Seneca: A Philosopher in Politics*, Oxford: Clarendon Press, rev. 1992.

Gross, N. (1989) *Senecas Naturales quaestiones: Komposition, naturphilosophische Aussagen und ihre Quellen*, Stuttgart: Steiner.

Gudger, E.W. (1924) 'Pliny's Historia Naturalis, the most popular natural history ever published', *Isis* 6: 269–81.

Guthrie, W.K.C. (1939) *Aristotle, On the Heavens*, Cambridge, Mass.: Harvard University Press (*LCL*).

—— (1962) *A History of Greek Philosophy*, vol. 1, *The Earlier Presocratics and the Pythagoreans*, Cambridge: Cambridge University Press.

Hadzsits, G.D. (1935) *Lucretius and his Influence*, London: George G. Harrap & Co.

Hahm, D.E. (1977) *The Origins of Stoic Cosmology*, [Columbus]: Ohio State University Press.

Hall, J.J. (1970) 'Ancient theories of wind, and the physical principles on which they are based, from the earliest times to Theophrastus', unpublished doctoral thesis, University of Cambridge.

—— (1977) 'Seneca as a source for earlier thought (especially meteorology)', *Classical Quarterly* (New Series) 27: 409–36.

Hammer-Jensen, I. (1915) 'Das Sogenannte IV. Buch der Meteorologie des Aristoteles', *Hermes* 50: 113–36.

Hankinson, R.J. (1995a) 'Philosophy of science', in Barnes, J. (ed.) *The Cambridge Companion to Aristotle*, Cambridge: Cambridge University Press, 109–39.

—— (1995b) 'Science', in Barnes, J. (ed.) *The Cambridge Companion to Aristotle*, Cambridge: Cambridge University Press, 140–67.

—— (1998) *Cause and Explanation in Ancient Greek Thought*, Oxford: Clarendon Press.

—— (1999) 'Explanation and causation', in Algra, K., Barnes, J., Mansfeld, J. and Schofield, M. (eds) *The Cambridge History of Hellenistic Philosophy*, Cambridge: Cambridge University Press, 479–512.

—— (forthcoming) 'Mathematics and physics in Aristotle's theory of the ether'.

Hannah, R. (1999) 'The Athenian star-calendar: tool of the state?', unpublished paper delivered to the 'On time: history, science and commemoration' conference, Liverpool, 1999, and also to the Australian Society for Classical Studies Conference, Adelaide, 2001.

—— (2001) 'From orality to literacy? The case of the parapegma', in *Speaking Volumes: Orality and Literacy in the Greek and Roman World*, Watson, J. (ed.) Leiden: Brill, 139–59.

—— (2002) 'Euctemon's parapegma', in Tuplin, C.J. and Rihll, T.E. (eds) *Science and Mathematics in Ancient Greek Culture*, Oxford: Oxford University Press, 112–32.

—— (forthcoming) *Greek and Roman Calendars*, London: Duckworth.

Hanson, A.E. (1991) 'Ancient illiteracy', in *Literacy in the Ancient World*, *Journal of Roman Archaeology*, Supplementary Series 3, 159–97.

Hardie, P.R. (1986) *Virgil's Aeneid: Cosmos and Imperium*, Oxford: Clarendon Press.

Harley, J.B. and Woodward, D. (1987) *The History of Cartography*, vol. 1, *Cartography in Prehistoric, Ancient, and Medieval Europe and the Mediterranean*, Chicago: University of Chicago Press.

Harris, W.V. (1989) *Ancient Literacy*, Cambridge, Mass.: Harvard University Press.

Healy, J.F. (1999) *Pliny the Elder on Science and Technology*, Oxford: Oxford University Press.

Heath, T. (1913) *Aristarchus of Samos: The Ancient Copernicus*, Oxford: Clarendon Press; reprinted 1981, New York: Dover.

—— (1949) *Mathematics in Aristotle*, Oxford: Clarendon Press.

Hellmann, G. (1916) 'Über die ägyptischen Witterungsangaben im Kalender von Claudius Ptolemaeus', *Sitzungsberichte der königlich Preussischen Akademie der Wissenschaften* 1: 332–41.

—— (1917) *Beiträge zur Geschichte der Meteorologie*, nr. 7: 'Die Witterungsangaben in den griechischen und lateinischen Kalendern', *Veröffentlichungen des Königlich Preußischen Meteorologischen Instituts* 296 (Berlin): 137–66.

Heninger, S.K. Jr (1960) *A Handbook of Renaissance Meteorology*, Durham, NC: Duke University Press.

Hesse, M.B. (1966) *Models and Analogies in Science*, Notre Dame, Indiana: University of Notre Dame.

Hett, W.S. (1936–1937) *Aristotle: Problems*, 2 vols, Cambridge, Mass.: Harvard University Press (*LCL*).

—— (1936) *Aristotle: Minor Works*, London: William Heinemann (*LCL*).

Hine, H.M. (1981) *An Edition with Commentary of Seneca, Natural Questions, Book Two*, New York: Arno Press.

—— (1995) 'Seneca's *Natural Questions* – changing readerships', in Ayres, L. (ed.) *The Passionate Intellect: Essays on the Transformation of Classical Traditions, presented to Professor I.G. Kidd*, New Brunswick: Transaction, *Rutgers University Studies in Classical Humanities*, vol. 7, 203–11.

—— (1996) *L. Annaei Senecae Naturalium Quaestionum Libros*, Stuttgart: Teubner.

—— (2002) 'Seismology and vulcanology in antiquity', in Tuplin, C.J. and Rihll, T.E. (eds) *Science and Mathematics in Ancient Greek Culture*, Oxford: Oxford University Press, 56–75.

Hornblower, S. and Spawforth, A. (eds) (1996) *The Oxford Classical Dictionary*, 3rd edn, Oxford: Oxford University Press.

Hort, A. (1916–26) *On Weather Signs* in *Theophrastus: Enquiry into Plants and Minor Works, On Odours and Weather Signs*, 2 vols, Cambridge, Mass.: Harvard University Press (*LCL*), vol. 2.

Housman, A.E. (1900) 'The Aratea of Germanicus', *Classical Review* 14: 26–39; reprinted in Housman, A.E. (1972) *The Classical Papers of A.E. Housman*, (eds) Diggle, J. and Goodyear, F.R.D., 3 vols, Cambridge: Cambridge University Press, vol. 2, 495–515.

Huffman, C.A. (1993) *Philolaus of Croton: Pythagorean and Presocratic*, Cambridge: Cambridge University Press.

Hunger, H. (1976) 'Astrologische Wettervorhersagen', *Zeitschrift für Assyriologie* 66: 234–60.

Hunter, R. (1995) 'Written in the stars: poetry and philosophy in the *Phainomena* of Aratus', *Arachnion* 2 (September): 1–34.

Hussey, Edward (1991a) 'Aristotle on mathematical objects', in Mueller, I. (ed.) *Peri Tōn Mathēmatōn, Apeiron* 24: 105–33.

—— (1991b) 'Aristotle's mathematical physics: a reconstruction', in Judson, L. (ed.) *Aristotle's* Physics: *A Collection of Essays*, Oxford: Clarendon Press, 213–42.

Ideler, J.L. (1832) *Meteorologia veterum Graecorum et Romanorum: prolegomena ad novam meteorologicorum Aristotelis*, Berlin: G.C. Nauck.

—— (1834–36) *Aristotelis Meteorologicorum libri IV*, 2 vols, Leipzig: F.C.G. Vogel.

Ideler, L. (1816) 'Über den Kalender des Ptolemäus', *Abhandlungen der Königlichen Akademie der Wissenschaften aus den Jahren 1816–1817*, Berlin, 163–214.

Ierodiakonou, K. (1995) 'Alexander of Aphrodisias on medicine as a stochastic art', in Eijk, P.J. van der, Horstmanshoff, H.F.J. and Schrijvers, P.H. (eds) *Ancient Medicine in its Socio-Cultural Context*, 2 vols, Amsterdam: Rodopi, *Clio Medica* 28, vol. 2, 473–85.

Illinois Greek Club (1923) *Aeneas Tacticus Asclepiodotus Onasander*, London: William Heinemann (*LCL*).

Inwards, R. (1893) *Weather Lore: A Collection of Proverbs, Sayings, and Rules concerning the Weather*, London: Elliot Stock.

Inwood, B. (1995) 'Seneca in his philosophical milieu', *Harvard Studies in Classical Philology* 97: 63–76.

—— (1999) 'God and human knowledge in Seneca's *Natural Questions*', *Proceedings of the Boston Area Colloquium in Ancient Philosophy* 15: 23–43.

—— (2002) 'God and human knowledge in Seneca's *Natural Questions*' (expanded version of Inwood 1999), in Frede, D. and Laks, A. (eds) *Traditions of Theology: Studies in Hellenistic Theology, its Background and Aftermath* (proceedings of the 1998 Symposium Hellenisticum), Leiden: Brill, 119–57.

Isager, S. and Skydsgaard, J.E. (1992) *Ancient Greek Agriculture: An Introduction*, London: Routledge.

Jackson, H. (1920) 'Aristotle's lecture-room and lectures', *Journal of Philology* 35: 191–200.

James, A.W. (1972) 'The Zeus hymns of Cleanthes and Aratus', *Antichthon* 6: 28–38.

Janković, V. (1998) 'Meteors under scrutiny: private, public, and professional weather in Britain, 1660–1800', unpublished doctoral thesis, Notre Dame University.

—— (2000) *Reading the Skies: A Cultural History of English Weather*, Chicago: University of Chicago Press.

Jenks, S. (1983) 'Astrometeorology in the Middle Ages', *Isis* 74: 185–210.

Jermyn, L.A.S. (1951) 'Weather-signs in Virgil', *Greece & Rome* 20: 26–37 and 49–59.

Jones, A. (1991) 'The adaptation of Babylonian methods in Greek numerical astronomy', *Isis* 82: 441–53.

—— (1999) 'Geminus and the Isia,' *Harvard Studies in Classical Philology* 99: 255–67.

Judge, E.A. (1997) 'The rhetoric of inscriptions', in Porter, S.E. (ed.) *Handbook of Classical Rhetoric in the Hellenistic Period 330 B.C.–A.D. 400*, Leiden: Brill, 807–28.

Judson, L. (ed.) (1991a) *Aristotle's* Physics: *A Collection of Essays*, Oxford: Clarendon Press.

—— (1991b) 'Chance and "always or for the most part" in Aristotle', in Judson, L. (ed.) *Aristotle's* Physics: *A Collection of Essays*, Oxford: Clarendon, 73–99.

Kahn, C.H. (1960) *Anaximander and the Origins of Greek Cosmology*, New York: Columbia University Press.

—— (1997) 'Greek religion and philosophy in the Sisyphus fragment', *Phronesis* 42: 47–62.

Kanelopoulos, C. (1990) 'L'agriculture d'Hésiode', *Techniques et Culture* 15: 131–58.

Khrgian, A.Kh. (1959) *Meteorology: A Historical Survey*, 2nd edn, trans. from Russian (*Ocherki razvitiya meteorologii*), ed. Kh.P. Pogosyan, Jerusalem: Israel Program for Scientific Translations, 1970.

Kidd, D. (1997) *Aratus: Phaenomena*, Cambridge: Cambridge University Press.

Kidd, I.G. (1978) 'Philosophy and Science in Posidonius', *Antike und Abendland* 24: 7–15.

—— (1988) *Posidonius II. The Commentary*, 2 vols, Cambridge: Cambridge University Press.

—— (1992) 'Theophrastus' *Meteorology*, Aristotle and Posidonius', in Fortenbaugh, W.W. and Gutas, D. (eds) *Theophrastus: His Psychological, Doxographical and Scientific Writings*, New Brunswick: Transaction, *Rutgers University Studies in Classical Humanities*, vol. 5, 294–306.

—— (1996) 'Posidonius', in Hornblower, S. and Spawforth, A. (eds) *The Oxford Classical Dictionary*, 3rd edn, Oxford: Oxford University Press.

—— (1997) 'What is a Posidonian fragment?', in Most, G. (ed.) *Collecting Fragments, Fragmente sammeln, Aporemata: Kritische Studien zur Philologiegeschichte* 1, Göttingen: Vandenhoeck & Ruprecht, 225–36.

—— (1999) *Posidonius III: The Translation of the Fragments*, Cambridge: Cambridge University Press.

Kirchner, J. (1931) *Inscriptiones Graecae*, vols II & III (Editio minor, pars altera), *Inscriptiones Atticae Euclidis Anno Posteriores* 2.2, Berlin: Walter de Gruyter.

Kirk, G.S., Raven, J.E. and Schofield, M. (1983) *The Presocratic Philosophers*, 2nd edn, Cambridge: Cambridge University Press.

Klein, S. (2000) *Endoxic Method and Ethical Inquiry: An Analysis and Defense of a Method for Justifying Fundamental Ethical Principles*, New York: P. Lang.

Kroll, W. (1951) 'C. Plinius Secundus der Ältere', in *Paulys Real-Encyclopädie der classischen Altertumswissenschaft* 21.1, columns 430–9.

Kullmann, W. (1998) 'Zoologische Sammelwerke in der Antike', in Kullmann, W., Althoff, J. and Asper, M. (eds) *Gattungen wissenschaftlicher Literatur in der Antike*, Tübingen: Gunter Narr Verlag, *ScriptOralia* 95, 121–39.

Kusukawa, S. (2000) 'Incunables and sixteenth-century books', in Hunter, A. (ed.) *Thornton and Tully's Scientific Books, Libraries and Collectors*, 4th edn, Aldershot: Ashgate.

Lais, P.G. (1894) 'Monumento Greco-Latino di una rosa classica dodecimale in Vaticano', *Pubblicazioni della Specola Vaticana* 4: xi–xv and plate I.

Laks, A. (1983) *Diogène d'Apollonie, La dernière cosmologie présocratique*, Lille: Presses Universitaires de Lille.

Laks, A. and Most, G.W. (1993) *Théophraste: Métaphysique*, Paris: Les Belles Lettres.

Laks, A., Most, G.W. and Rudolph, E. (1988) 'Four notes on Theophrastus' metaphysics', in Fortenbaugh, W.W. and Sharples, R.W. (eds) *Theophrastean Studies: On Natural Science, Physics and Metaphysics, Ethics, Religion, and Rhetoric*, New Brunswick: Transaction, *Rutgers University Studies in Classical Humanities*, vol. 3, 224–56.

Lamberton, R. (1988) *Hesiod*, New Haven: Yale University Press.

Lamberton, R. and Keaney, J.J. (eds) (1992) *Homer's Ancient Readers: The Hermeneutics of Greek Epic's Earliest Exegetes*, Princeton: Princeton University Press.

Langslow, D.R. (1999) 'The language of poetry and the language of science: the Latin poets and "medical Latin"', *Proceedings of the British Academy* 93: 183–225.

Lear, J. (1982) 'Aristotle's philosophy of mathematics', *The Philosophical Review* 91: 161–92.

Lee, H.D.P. (1952) *Aristotle, Meteorologica*, Cambridge, Mass.: Harvard University Press (*LCL*).

Lehoux, D. (2000) 'Parapegmata, or astrology, weather, and calendars in the ancient world, being an examination of the interplay between the heavens and the earth in the classical and near-eastern cultures of antiquity, with particular reference to the regulation of agricultural practice, and to the signs and causes of storms, tempests, &c.', unpublished doctoral thesis, University of Toronto.

Lennox, J.G. (1985) 'Theophrastus on the limits of teleology', in Fortenbaugh, W.W., Huby, P.M. and Long, A.A. (1985) *Theophrastus of Eresus: On His Life and Work*, New Brunswick: Transaction, *Rutgers University Studies in Classical Humanities*, vol. 2, 143–63.

—— (2001) *Aristotle's Philosophy of Biology: Studies in the Origins of Life Science*, Cambridge: Cambridge University Press.

Lettinck, P. (1999) *Aristotle's Meteorology and its Reception in the Arab World*, Leiden: E.J. Brill.

Lewis, E. (1996) *Alexander of Aphrodisias: On Aristotle Meteorology 4*, London: Duckworth.

Lewis, G.C. (1862) *An Historical Survey of the Astronomy of the Ancients*, London: Parker, Son, and Bourn.

Liddell, H.G., Scott, R. and Jones, H.S. (1940) *A Greek–English Lexicon*, 9th edn, Oxford: Clarendon Press, with *Supplement* by Barber, E.A. *et al.*, 1968, reprinted in 1 vol.

Linderski, J. (1996) 'Divination', in Hornblower, S. and Spawforth, A. (eds) *The Oxford Classical Dictionary*, 3rd edn, Oxford: Oxford University Press.

Lippincott, K., Eco, U., Gombrich, E.H. and others (1999) *The Story of Time*, London: Merrell Holberton.

Lipton, Peter (1991) *Inference to the Best Explanation*, London: Routledge.

Lloyd, G.E.R. (1964) 'Experiment in early Greek philosophy and medicine', *Proceedings of the Cambridge Philological Society* N.S. 10 (1964), 50–72; reprinted in Lloyd, G.E.R. (1991) *Methods and Problems in Greek Science: Selected Papers*, Cambridge: Cambridge University Press, 70–99.

—— (1966) *Polarity and Analogy: Two Types of Argumentation in Early Greek Thought*, Cambridge: Cambridge University Press; reprinted 1987 and 1992 by Bristol Classical Press and Hackett Publishing.

—— (1968) *Aristotle: The Growth and Structure of his Thought*, Cambridge: Cambridge University Press.

—— (1970) *Early Greek Science: Thales to Aristotle*, New York: W.W. Norton.

—— (1978) 'The empirical basis of the physiology of the *Parva Naturalia*', in Lloyd, G.E.R. and Owen, G.E.L., *Aristotle on Mind and the Senses*, Cambridge: Cambridge University Press, 215–39; reprinted in Lloyd (1991), 224–47.

—— (1979) *Magic, Reason and Experience: Studies in the Origin and Development of Greek Science*, Cambridge: Cambridge University Press.

—— (1987) *The Revolutions of Wisdom: Studies in the Claims and Practice of Ancient Greek Science*, Cambridge: Cambridge University Press.

—— (1991) *Methods and Problems in Greek Science*, Cambridge: Cambridge University Press.

—— (1996a) *Adversaries and Authorities: Investigations into Ancient Greek and Chinese Science*, Cambridge: Cambridge University Press.

—— (1996b) *Aristotelian Explorations*, Cambridge: Cambridge University Press.

Long, A.A. (1982) 'Astrology: arguments pro and contra', in Barnes, J., Brunschwig, J., Burnyeat, M. and Schofield, M. (eds) *Science and Speculation: Studies in Hellenistic Theory and Practice*, Cambridge: Cambridge University Press, 165–92.

—— (1992) 'Stoic readings of Homer', in Lamberton, R. and Keaney, J.J. (eds) *Homer's Ancient Readers: The Hermeneutics of Greek Epic's Earliest Exegetes*, Princeton: Princeton University Press, 41–66.

Long, A.A. and Sedley, D.N. (1987) *The Hellenistic Philosophers*, 2 vols, Cambridge: Cambridge University Press.

Longrigg, J. (1966) Review of P. Steinmetz (1964) *Die Physik des Theophrastos von Eresos*, Bad Homburg: Dr Max Gehlen, in *Classical Review* (New Series) 16: 177–9.

—— (1975) 'Elementary physics in the Lyceum and Stoa', *Isis* 66: 211–29.

Loose, G. (1995) 'Das 2. Buch der Naturalis Historia von Plinius dem Älteren: eine kritische Analyse im Lichte moderner geowissenschaftlicher Erkenntnisse', Ph.D. thesis (1993), University of Cologne, Cologne: P & P.

Ludlum, D.M. (1991, 2001) *Collins Wildlife Trust Guide to the Weather of Britain and Europe: A Photographic Guide to the Weather of Britain and Europe*, London: Harper Collins.

Maas, M. (1992) *John Lydus and the Roman Past*, London: Routledge.

Maass, E. (1898) *Commentariorum in Aratum Reliquiae*, Berlin: Weidmann.

McLeod, O. (1995) 'Aristotle's method', *History of Philosophy Quarterly* 12: 1–18.

Madden, E.H. (1957) 'Aristotle's treatment of probability and signs', *Philosophy of Science* 24: 167–72.

Maguire, J.P. (1939) 'The sources of the pseudo-Aristotle De Mundo', *Yale Classical Studies* 6: 111–67.

Mair, A.W. (1908) *Hesiod: The Poems and Fragments*, Oxford: Clarendon Press.

Mair, G.R. (1921) *Aratus: Phaenomena*, in *Callimachus, Lycophron, Aratus*, Cambridge, Mass.: Harvard University Press (*LCL*).

Manitius, C. (1898) *Gemini Elementa astronomiae*, Stuttgart: Teubner.

Mansfeld, J. (1971) *The Pseudo-Hippocratic Tract* ΠΕΡΙ ΕΒΔΟΜΑΔΩΝ *Ch. 1–11 and Greek Philosophy*, Assen: Van Gorcum.

—— (1990) 'Doxography and dialectic. The *Sitz in Leben* of the "Placita"', in *Aufstieg und Niedergang der römischen Welt (ANRW)* 2.36.4: 3056–229.

—— (1992a) '*Physikai Doxai* and *Problēmata physika* from Aristotle to Aëtius (and beyond)', in Fortenbaugh, W.W. and Gutas, D. (1992) *Theophrastus: His Psychological, Doxographical and Scientific Writings*, New Brunswick: Transaction, *Rutgers University Studies in Classical Humanities*, vol. 5, 63–111.

—— (1992b) 'A Theophrastean excursus on god and nature and its aftermath in Hellenistic thought', *Phronesis* 37: 314–35.

—— (1992c) 'ΠΕΡΙ ΚΟΣΜΟΥ: A note on the history of a title', *Vigiliae Christianae* 46: 391–411.

—— (1998) 'Doxographical studies, Quellenforschung, tabular presentation and other varieties of comparativism', in Burkert, W., Marciano, L.G., Matelli, E. and Orelli, L. (eds) *Fragmentsammlungen philosophischer Texte der Antike, Le raccolte dei frammenti di filosofi antichi, Atti del Seminario Internazionale Ascona, Centro Stefano Franscini 22–27 Settembre 1996, Aporemata 3*, Göttingen: Vandenhoeck & Ruprecht, 16–40.

—— (1999a) 'Sources', in Algra, K., Barnes, J., Mansfeld, J. and Schofield, M. (eds) *The Cambridge History of Hellenistic Philosophy*, Cambridge: Cambridge University Press, 3–30.

—— (1999b) 'Theology', in Algra, K., Barnes, J., Mansfeld, J. and Schofield, M. (eds) *The Cambridge History of Hellenistic Philosophy*, Cambridge: Cambridge University Press, 452–78.

Mansfeld, J. and Runia, D. (1997) *Aëtiana: The Method and Intellectual Context of a Doxographer*, vol. 1, *The Sources*, Leiden: Brill.

Marsilio, M.S. (2000) *Farming and Poetry in Hesiod's Works and Days*, Lanham: University Press of America.

Matthen, M. (2001) 'The holistic presuppositions of Aristotle's cosmology', *Oxford Studies in Ancient Philosophy* 20: 171–99.

Maurach, G. (ed.) (1975a) *Seneca als Philosoph*, Darmstadt: Wissenschaftliche Buchgesellschaft.

—— (1975b) 'Zur Eigenart und Herkunft von Senecas Methode in den "Naturales Quaestiones"', in Maurach, G. (ed.) *Seneca als Philosoph*, Darmstadt: Wissenschaftliche Buchgesellschaft, 305–22.

Maxwell-Stuart, P.G. (1996) 'Theophrastus the traveller', *La Parola del Passato: Rivista di Studi Antichi* 289: 241–67.

Meritt, B.D. (1928) *The Athenian Calendar in the Fifth Century*, Cambridge, Mass.: Harvard University Press.

Mingazzini, P. (1928) 'VI. Pozzuoli. Frammento di calendario perpetuo', *Notizie degli Scavi di Antichità* 6.4: 202–5.

Mitsis, P. (1993) 'Committing philosophy on the reader: didactic coercion and reader autonomy in De Rerum Natura', in Schiesaro, A., Mitsis, P. and Clay, J.S. (eds) *Mega Nepios: il destinatario nell'epos didascalico*, Pisa: Giardini, *Materiali e discussioni per l'analisi dei testi classici* 31, 111–28.

Modrak, D.K.W. (1989) 'Aristotle on the difference between mathematics and physics and first philosophy', in Penner, T. and Kraut, R. (eds) *Nature, Knowledge and Virtue: Essays in Memory of Joan Kung, Apeiron* 23: 121–39.

Monmonier, M. (2000) *Air Apparent: How Meteorologists Learned to Map, Predict and Dramatize Weather*, Chicago: University of Chicago Press.

Mourelatos, A.P.D. (1989) ' "X is really Y": Ionian origins of a thought pattern', in Boudouris, K.J. (ed.) *Ionian Philosophy*, Athens: International Association for Greek Philosophy, 280–90.

Mueller, I. (1970) 'Aristotle on geometrical objects', *Archiv für Geschichte der Philosophie* 52.19: 156–71.

Müller, C. (1861) *Geographi Graeci Minores*, 2 vols, Paris: Firmin Didot.

Nauert, C.G. Jr (1980) 'Caius Plinius Secundus', *Catalogus Translationum et Commentariorum* 4: 297–422.

Nelson, S.A. (1998) *God and the Land: The Metaphysics of Farming in Hesiod and Vergil*, New York and Oxford: Oxford University Press.

Netz, Reviel (1998) 'The first Jewish scientist?', *Scripta Classica Israelica* 17: 27–33.

—— (1999) *The Shaping of Deduction in Greek Mathematics: A Study in Cognitive History*, Cambridge: Cambridge University Press.

Neuberg, M. (1993) 'Hitch your wagon to a star: Manilius and his two addressees', in Schiesaro, A., Mitsis, P. and Clay, J.S. (eds) *Mega Nepios: il destinatario nell'epos didascalico*, Pisa: Giardini, *Materiali e discussioni per l'analisi dei testi classici* 31, 243–82.

Neugebauer, O. (1975) *A History of Ancient Mathematical Astronomy*, 3 vols, Berlin: Springer.

Nikitinski, O. (1998) 'Plinius der Ältere: Seine Enzyklopädie und ihre Leser', in Kullmann, W., Althoff, J. and Asper, M. (eds) *Gattungen wissenschaftlicher Literatur in der Antike*, Tübingen: Gunter Narr Verlag, *ScriptOralia* 95, 341–59.

Noble, J.V. and Price, D.J. de S. (1968) 'The water clock in the tower of the winds', *American Journal of Archaeology* 72: 345–55.

North, J. (1994) *The Fontana History of Astronomy annd Cosmology*, London: Fontana Press.

Nussbaum, M. (1987) 'Saving Aristotle's appearances', in Schofield, M. and Nussbaum, M.C. (eds) (1987) *Language and Logos: Studies in Ancient Greek Philosophy presented to G.E.L. Owen*, Cambridge: Cambridge University Press, 267–93.

Obrist, B. (1997) 'Wind diagrams and medieval cosmology', *Speculum* 72: 33–84.

Ophuijsen, J.M. van and Raalte, M. van (eds) (1998) *Theophrastus: Reappraising the Sources*, New Brunswick: Transaction, *Rutgers University Studies in Classical Humanities*, vol. 8.

Owen, G.E.L. (1961) 'Tithenai ta phainomena', in Mansion, S. (ed.) *Aristote et les problèmes de méthode*, Louvain: Publications Universitaires de Louvain, 83–103; reprinted in Barnes, J., Schofield, M. and Sorabji, R. (eds.) *Articles on Aristotle*, 1, *Science*, London: Duckworth, 113–26.

Pauly, A., Wissowa, G. and Kroll, W. (1894–1980) (eds) *Paulys Real-Encyclopädie der classischen Altertumswissenschaft*, Stuttgart: J.B. Metzlerscher Verlag.

Pedersen, Olaf (1974) *A Survey of the Almagest*, [Odense]: Odense University Press.

—— (1986) 'Some astronomical topics in Pliny', in French, R. and Greenaway, F. (eds) *Science in the Early Roman Empire: Pliny the Elder, his Sources and Influence*, London: Croom Helm, 162–96.

Pfeiffer, E. (1916) *Studien zum antiken Sternglauben, Stoicheia* Heft 2, Leipzig: B.G. Teubner.

Pinsent, J. (1989) 'Boeotian calendar poetry', in Beister, H. and Buckler, J. (eds) *Boiotika: Vorträge vom 5. Internationalen Böotien-Kolloquium zu Ehren von Professor Dr. Siegried Lauffer*, Institut für Alte Geschichte Ludwig-Maximilians-Universität München 13–17 Juni 1986, Munich: Editio Maris, *Münchener Arbeiten zur alten Geschichte* 2: 33–7.

Pöhlmann, E. (1973) 'Charakteristika des römischen Lehrgedichts', in *Aufstieg und Niedergang der römischen Welt (ANRW)* I.3: 814–901.

Preus, A. (1990) 'Man and cosmos in Aristotle: metaphysics Λ and the biological works', in Devereux, D. and Pellegrin, P. (eds) *Biologie, Logique et Métaphysique chez Aristote: Actes du Séminaire C.N.R.S.-N.S.F. Oléron 28 juin–3 juillet 1987*, Paris: Éditions du CNRS.

Price, D. de S. (1974) *Gears from the Greeks: The Antikythera Mechanism – A Calendar Computer from ca. 80 B.C.*, *Transactions of the American Philosophical Society (New Series)* 64, part 7.

Pritchett, W.K. and Waerden, B.L. van der (1961) 'Thucydidean time-reckoning and Euctemon's seasonal calendar', *Bulletin de Correspondance Hellénique* 85: 17–52.

Pritzl, K. (1994) 'Opinions as appearances: *endoxa* in Aristotle', *Ancient Philosophy* 14: 41–50.

Purcell, N. (1996) 'Pliny the Elder', in Hornblower, S. and Spawforth, A. (eds) *The Oxford Classical Dictionary*, 3rd edn, Oxford: Oxford University Press.

Raalte, M. van (1988) 'The idea of the cosmos as an organic whole in Theophrastus' Metaphysics', in Fortenbaugh, W.W. and Sharples, R.W. (eds) *Theophrastean Studies: On Natural Science, Physics and Metaphysics, Ethics, Religion, and Rhetoric*, New Brunswick: Transaction, *Rutgers University Studies in Classical Humanities,* vol. 3, 189–215.

—— (1993) *Theophrastus: Metaphysics*, Leiden: E.J. Brill.

—— (forthcoming) 'God and the nature of the world: the "theological excursus" in Theophrastus' *Meteorology*', *Mnemosyne*.

Rackham, H. *et al.* (1938–63) *Pliny: Natural History*, 10 vols, Cambridge, Mass.: Harvard University Press (*LCL*).

Reale, G. and Bos, A.P. (1995) *It trattato* Sul cosmo per Alessandro *attribuito ad Aristotele*, 2nd edn, Milan: Vita e Pensiero.

Regenbogen, O. (1940) 'Theophrastos', *Paulys Real-Encyclopädie der classischen Altertumswissenschaft*, Supplementband 7, columns 1354–562.

Rehm, A. (1904) 'Weiteres zu den milesischen Parapegmen', *Sitzungsberichte der Königlich Preussischen Akademie der Wissenschaften, philosophisch-historischen Classe* 23: 752–59.

—— (1913) 'III. Das Parapegma des Euktemon', in Boll, F. (ed.) *Griechische Kalender*, in *Sitzungsberichte der Heidelberger Akademie der Wissenschaften, philosophisch-historische Klasse*, Abhandlung 3: 3–38.

—— (1916) *Griechische Windrosen, Sitzungsberichte der königlich Bayerischen Akademie der Wissenschaften, philosophisch-philologische und historische Klasse*, Abhandlung 3: 1–104.

—— (1940) 'Episemasiai', in Pauly, A., Wissowa, G. and Kroll, W. (eds) *Paulys Real-Encyclopädie der classischen Altertumswissenschaft*, Supplement-band 7, columns 175–98.

—— (1941) *Parapegmastudien, Abhandlungen der Bayerischen Akademie der Wissenschaften, philosophisch-historische Abteilung Neue Folge* 19: 1–145.

—— (1949) 'Parapegma', in Pauly, A., Wissowa, G. and Kroll, W. (eds) *Paulys Real-Encyclopädie der classischen Altertumswissenschaft* 18.4, columns 1295–366.

—— (1975) 'Das siebente Buch der *Naturales Quaestiones* des Seneca und die Kometentheorie des Poseidonios', in Maurach, G. (ed.) (1975) *Seneca als Philosoph*, Darmstadt: Wissenschaftliche Buchgesellschaft, 228–63.

Renon, L.V. (1998) 'Aristotle's *endoxa* and plausible argumentation', *Argumentation* 12: 95–113.

Repici, L. (1990) 'Limits of teleology in Theophrastus' Metaphysics?', *Archiv für Geschichte der Philosophie* 72: 182–213.

Richardson, N.J. (1992) 'Aristotle's reading of Homer', in Lamberton, R. and Keaney, J.J. (eds) *Homer's Ancient Readers: The Hermeneutics of Greek Epic's Earliest Exegetes*, Princeton: Princeton University Press, 30–40.

Robinson, H.S. (1943) 'The tower of the winds and the Roman market-place', *Supplement to the American Journal of Archaeology* 47: 291–305.

Röhr, J. (1928) 'Beiträge zur antiken Astrometeorologie', *Philologus* 83: 259–305.

Ross, A. (1991) *Strange Weather: Culture, Science and Technology in the Age of Limits*, New York: Verso.

Ross, D.O. Jr (1986) *Virgil's Elements: Physics and Poetry in the* Georgics, Princeton: Princeton University Press.

Ross, W.D. (1923) *Aristotle*, London: Methuen.

—— (1924) *Aristotle: Metaphysics*, Oxford: Clarendon Press.

—— (1936) *Aristotle: Physics*, Oxford: Clarendon Press.

Ross, W.D. and Fobes, F.H. (1929) *Theophrastus: Metaphysics*, Oxford: Clarendon Press.

Runia, D. (1989) 'Xenophanes on the moon: a *doxographicum* in Aëtius', *Phronesis* 34/3: 245–69.

Sale, W. (1972) 'The Olympian faith', *Greece and Rome* (Second Series) 19: 81–93.

Sallares, R. (1991) *The Ecology of the Ancient Greek World*, London: Duckworth.

Sambursky, S. (1956a) 'On the possible and the probable in ancient Greece', *Osiris* 12: 35–48.

—— (1956b) *The Physical World of the Greeks*, trans. M. Dagut, Princeton: Princeton University Press.

—— (1959) *Physics of the Stoics*, Princeton: Princeton University Press.

Sayili, A.M. (1939) 'The Aristotelian explanation of the rainbow', *Isis* 30: 65–83.

Schechner Genuth, S. (1997) *Comets, Popular Culture, and the Birth of Modern Cosmology*, Princeton: Princeton University Press.

Schenkeveld, D.M. (1991) 'Language and style of the Aristotelian *De mundo* in relation to the question of its inauthenticity', in *Elenchos* 12: 221–55.

—— (1997) 'Philosophical prose', in Porter, S.E. (ed.) *Handbook of Classical Rhetoric in the Hellenistic Period 330 B.C.–A.D. 400*, Leiden: Brill, 195–264.

Schiesaro, A. (1996) 'Didactic poetry', in Hornblower, S. and Spawforth, A. (eds) *The Oxford Classical Dictionary*, 3rd edn, Oxford: Oxford University Press.

Schiesaro, A., Mitsis, P. and Clay, J.S. (eds) (1993) *Mega Nepios: il destinatario nell'epos didascalico*, Pisa: Giardini, *Materiali e discussioni per l'analisi dei testi classici* 31.

Schofield, M. (1986) 'Cicero for and against divination', *Journal of Roman Studies* 76: 47–65.

Schofield, M. and Nussbaum, M.C. (eds) (1982) *Language and Logos: Studies in Ancient Greek Philosophy presented to G.E.L. Owen*, Cambridge: Cambridge University Press.

Scholfield, A.F. (1958) *Aelian: On the Characteristics of Animals*, 3 vols, Cambridge, Mass.: Harvard University Press (*LCL*).

Schoonheim, P.L. (2000) *Aristotle's Meteorology in the Arabico-Latin Tradition: A Critical Edition of the Texts*, Leiden: Brill.

Scullard, H.H. (1981) *Festivals and Ceremonies of the Roman Republic*, London: Thames and Hudson.

Seabrook, J. (2000) 'Selling the weather: a national obsession for dire forecasts', *The New Yorker*, 3 April 2000: 44–53.

Sedley, D. (1982) 'On signs', in Barnes, J., Brunschwig, J., Burnyeat, M. and Schofield, M. (eds) (1982) *Science and Speculation: Studies in Hellenistic Theory and Practice*, Cambridge: Cambridge University Press, 239–72.

—— (1991) 'Is Aristotle's teleology anthropocentric?', *Phronesis* 36/2: 179–96.

—— (1997) ' "Becoming like god" in the *Timaeus* and Aristotle', in Calvo, T. and Brisson, L. (eds) *Interpreting the* 'Timaeus-Critias': *Proceedings of the IV Symposium Platonicum Selected Papers*, Sankt Augustin: Academia Verlag, 327–39.

—— (1998a) *Lucretius and the Transformation of Greek Wisdom*, Cambridge: Cambridge University Press.

—— (1998b) 'Theophrastus and Epicurean physics', in Ophuijsen, J.M. van and Raalte, M. van (eds) (1998) *Theophrastus: Reappraising the Sources*, New Brunswick: Transaction, *Rutgers University Studies in Classical Humanities*, vol. 8, 331–54.

Seymour, T.D. (1907) *Life in the Homeric Age*, New York: Macmillan.

Sharples, R.W. (1987) 'Alexander of Aphrodisias: scholasticism and innovation', in *Aufstieg und Niedergang der römischen Welt (ANRW)* 2.36.2: 1176–83.

—— (1996a) *Stoics, Epicureans and Sceptics: An Introduction to Hellenistic Philosophy*, London: Routledge.

—— (1996b) 'Theophrastus', in Hornblower, S. and Spawforth, A. (eds) *The Oxford Classical Dictionary*, 3rd edn, Oxford: Oxford University Press.

—— (1998a) *Theophrastus of Eresus: Sources for his Life, Writings, Thought and Influence. Commentary Volume 3.1, Sources on Physics (Texts 137–233)*, with contributions on the Arabic material by Dimitri Gutas, Leiden: Brill, *Philosophia Antiqua* 79.

—— (1998b) 'Theophrastus as philosopher and Aristotelian', in Ophuijsen, J.M. van and Raalte, M. van (eds) (1998) *Theophrastus: Reappraising the Sources*, New Brunswick: Transaction, *Rutgers University Studies in Classical Humanities*, vol. 8, 267–80.

Sider, D. (2002a) 'Demokritos on the weather', in Laks, A. (ed.) *Qu'est-ce que la Philosophie Présocratique?*, Lille: Presses Universitaires de Lille, 287–302.

—— (2002b) 'On *On Signs*', in Fortenbaugh, W.M. and Wörhle, G. (eds) *On the* Opuscula *of Theophrastus: Akten der 3. Tagung der Karl- und Gertrud-Abel-Stiftung vom 19.–23. Juli 1999 in Trier*, Stuttgart: Franz Steiner Verlag, *Philosophie der Antike* 14: 99–111.

—— (forthcoming) *On Signs* (with W. Brunschön).

Smith, R. (1989) *Aristotle: Prior Analytics*, Indianapolis: Hackett.

—— (1995) 'Logic', in Barnes, J. (ed.) *The Cambridge Companion to Aristotle*, Cambridge: Cambridge University Press, 27–65.

Solmsen, F. (1957) 'Review of *Aristotle, Meteorologica*', trans. H.D.P. Lee (1952) (*LCL*), in *Gnomon* 29: 131–4.

—— (1960) *Aristotle's System of the Physical World: A Comparison with his Predecessors*, Ithaca, NY: Cornell University Press, *Cornell Studies in Classical Philology* 33.

Sorabji, R. (1972) 'Aristotle, mathematics and colour', *Classical Quarterly* (New Series) 22: 293–308.

Soubiran, J. (1969) *Vitruve, de l'architecture livre 9*, Paris: Les Belles Lettres.

—— (1981) *Aviénus: Les Phénomènes d'Aratos*, Paris: Les Belles Lettres.

Staden, H. von (1975) 'Experiment and experience in Hellenistic medicine', *Bulletin of the Institute of Classical Studies* 22: 178–99.

—— (1997) 'Teleology and mechanism: Aristotelian biology and early Hellenistic medicine', in Kullman, W. and Föllinger, S. (eds) *Aristotelische Biologie: Intentionen, Methoden, Ergebnisse, Akten des Symposions über Aristoteles' Biologie vom 24–28 Juli 1995 in der Werner-Reimers-Stiftung in Bad Homburg*, Stuttgart: Franz Steiner Verlag, 183–208.

Stahl, G. (1964) 'Die *Naturales Quaestiones* Senecas: Ein Beitrag zum Spiritualisierungsprozeß der römischen Stoa', *Hermes* 92: 425–54; reprinted in Maurach, G. (ed.) (1975) *Seneca als Philosoph*, Darmstadt: Wissenschaftliche Buchgesellschaft, 264–304.

Steinmetz, P. (1964) *Die Physik des Theophrast*, Bad Homburg: Max Gehlen, *Palingenesia* 1.

Stokes, M.C. (1962) 'Hesiodic and Milesian cosmogonies – I', *Phronesis* 7: 1–37.

—— (1963) 'Hesiodic and Milesian cosmogonies – II', *Phronesis* 8: 1–34.

Strohm, H. (1935) *Untersuchungen zur Entwicklungsgeschichte der aristotelischen Meteorologie, Philologu*, Supplementband 28, Heft 1, Leipzig: Dieterich'sche Verlagsbuchhandlung.

—— (1937) 'Zur Meteorologie des Theophrast', *Philologus* 92: 249–68 and 403–28.

—— (1952) 'Studien zur Schrift von der Welt', *Museum Helveticum* 9: 137–75.

—— (1953) 'Theophrast und Poseidonios: Drei Interpretationen zur Meteorologie', *Hermes* 81: 278–95.

—— (1979) *Aristoteles Meteorologie: Über die Welt*, 2nd edn, Berlin: Akademie-Verlag.

—— (1983) 'Beobachtungen zum vierten Buch der aristotelischen Meteorologie', in Moraux, P. and Wiesner, J. (eds) *Zweifelhaftes im Corpus Aristotelicum, Studien zu einigen Dubia. Akten des 9. Symposium Aristotelicum (Berlin 7–16 September 1981)*, Berlin: Walter de Gruyter.

—— (1987) 'Ps. Aristoteles De Mundo und Theilers Poseidonios', *Wiener Studien* 100: 69–84.

Svenbro, J. (1988) *Phrasikleia: Anthropologie de la lecture en Grèce ancienne*, trans. J. Lloyd (1993) *Phrasikleia: An Anthropology of Reading in Ancient Greece*, Ithaca: Cornell University Press.

Swerdlow, N.M. (1998) *The Babylonian Theory of the Planets*, Princeton: Princeton University Press.

Tannery, P. (1893) *Recherches sur l'histoire de l'astronomie ancienne*, Paris: Gauthier-Villars & Fils.

Tate, J. (1927) 'The beginnings of Greek allegory', *Classical Review* 41: 214–15.

—— (1929) 'Cornutus and the poets', *Classical Quarterly* 23: 41–5.

—— (1929, 1930) 'Plato and allegorical interpretation', *Classical Quarterly* 23: 142–54 and 24: 1–10.

—— (1934) 'On the history of allegorism', *Classical Quarterly* 28: 105–14.

Taub, L. (1993) *Ptolemy's Universe: The Natural Philosophical and Ethical Foundations of Ptolemy's Astronomy*, Chicago: Open Court.

—— (1997) 'The rehabilitation of wretched subjects', *Early Science and Medicine* 2: 74–87.

—— (1998) 'Meteorology in the ancient world', in Good, G.A. (ed.) *Sciences of the Earth: An Historical Encyclopedia*, New York: Garland.

—— (2002a) 'Ancient meteorology: astronomy and weather prediction in the Roman period', in Renn, J. and Castagnetti, G. (eds) *Homo Faber: Studies on Nature, Technology and Science at the Time of Pompeii*, Rome: 'L'Erma' di Bretschneider, *Studi della Soprintendenza archeologica di Pompei* 6, 143–52.

—— (2002b) 'Instruments of Alexandrian astronomy: the uses of the equinoctial rings', in Tuplin, C.J. and Rihll, T.E. (eds) *Science and Mathematics in Ancient Greek Culture*, Oxford: Oxford University Press, 133–49.

Taylor, E.G.R. (1948) 'The navigator in antiquity', *Journal of the Institute of Navigation* 1: 103–8.

—— (1951) 'Early charts and the origin of the compass rose', *Journal of the Institute of Navigation* 4: 351–6.

Thalmann, W.G. (1984) *Conventions of Form and Thought in Early Greek Epic Poetry*, Baltimore: Johns Hopkins University Press.

Thomas, R. (1992) *Literacy and Orality in Ancient Greece*, Cambridge: Cambridge University Press.

Thomas, R.F. (1999) *Reading Virgil and His Texts: Studies in Intertextuality*, Ann Arbor: University of Michigan Press.

Thompson, D'A.W. (1910) *Historia Animalium*, in Smith, J.A. and Ross, W.D. (eds) *The Works of Aristotle Translated into English*, Vol. IV, Oxford: Clarendon Press.

—— (1918) 'The Greek winds', *Classical Review* 32: 49–56.

Timpanaro, S. (1994) 'Alcuni fraintendimenti del *De Divinatione*', in his *Nuovi Contributi di Filologia e Storia della Lingua Latina*, Bologna: Patron Editore, 241–64.

Toohey, P. (1996) *Epic Lessons: An Introduction to Ancient Didactic Poetry*, London: Routledge.

Toomer, G.J. (1974) 'Meton', in Gillespie, C.C. (ed.) (1970–80) *Dictionary of Scientific Biography*, New York: Charles Scribner's Sons, vol. 9.

—— (1984) *Ptolemy's Almagest*, New York: Springer Verlag.

—— (1988) 'Hipparchus and Babylonian astronomy', in Leichty, E., Ellis, M. de J. and Gerardi, P. (eds) *A Scientific Humanist: Studies in Memory of Abraham Sachs*, Philadelphia: Occasional Publications of the Samuel Noah Kramer Fund 9, 353–62.

—— (1996a) 'Aratus', in Hornblower, S. and Spawforth, A. (eds) *The Oxford Classical Dictionary*, 3rd edn, Oxford: Oxford University Press.

—— (1996b) 'Geminus', in Hornblower, S. and Spawforth, A. (eds) *The Oxford Classical Dictionary*, 3rd edn, Oxford: Oxford University Press.

—— (1996c) 'Meton', in Hornblower, S. and Spawforth, A. (eds) *The Oxford Classical Dictionary*, 3rd edn, Oxford: Oxford University Press.

Travlos, J. (1971) *Pictorial Dictionary of Ancient Athens*, prepared with the collaboration of the German Archaeological Institute, London: Thames and Hudson.

Tsagarakis, O. (1977) *Nature and Background of Major Concepts of Divine Power in Homer*, Amsterdam: B.R. Grüner.

Unger, G.F. (1869) 'Zu Ptolemäeus Φάσεις ἀπλανῶν', *Philologus* 28: 11–39.

Vallance, John (1988) 'Theophrastus and the study of the intractable', in Fortenbaugh, W.W. and Sharples, R.W. (eds) *Theophrastean Studies: On Natural Science, Physics, Metaphysics, Ethics, Religion and Rhetoric*, New Brunswick: Transaction, *Rutgers University Studies in Classical Humanities*, vol. 3, 25–40.

—— (1996) 'Meteorology', in Hornblower, S. and Spawforth, A. (eds) *The Oxford Classical Dictionary*, 3rd edn, Oxford: Oxford University Press.

—— (2001) 'Meteorologia', in *Storia della Scienza*, vol. 1: *La scienza antica*, *Instituto della Enciclopedia Italiana*, 863–73.

Vogt, H. (1920) 'Der Kalender des Claudius Ptolemäus', in Boll, F. (ed.) *Griechische Kalender, Sitzungsberichte der Heidelberger Akademie der Wissenschaften, philosophisch-historische Klasse* 15.

Volk, K. (2002) *The Poetics of Latin Didactic: Lucretius, Vergil, Ovid, Manilius*, Oxford: Oxford University Press.

Waerden, B.L. van der (1985) 'Greek Astronomical Calendars IV. The parapegma of the Egyptians and their "Perpetual Tables"', *Archive for History of Exact Sciences* 32: 95–104.

—— (1988) *Die Astronomie der Griechen*, Darmstadt: Wissenschaftliche Buchgesellschaft.

Waiblinger, F.P. (1977) *Senecas Naturales Quaestiones: Griechische Wissenschaft und römische Form*, Munich: C.H. Beck'sche Verlagsbuchhandlung, *Zetemata Monographien Heft* 70.

Wardy, R. (1993) 'Aristotelian rainfall or the lore of averages', *Phronesis* 38/1: 18–30.

Webster, E.W. (1931) '*Meteorologica*', in Ross, W.D. (ed.) *The Works of Aristotle*, Oxford: Clarendon Press, vol. 3.

Weidemann, H. (1989) 'Aristotle on inferences from signs', *Phronesis* 34: 342–51.

Wenskus, O. (1990) *Astronomische Zeitangaben von Homer bis Theophrast, Hermes*, Einzelschriften Heft 55, Stuttgart: Franz Steiner.

—— (1998) 'Columellas Bauernkalender zwischen Mündlichkeit und Schriftlichkeit', in Kullmann, W., Althoff, J. and Asper, M. (eds) *Gattungen wissenschaftlicher Literatur in der Antike*, Tübingen: Gunter Narr Verlag, ScriptOralia 95, 253–62.

Wessely, C. (1900) 'Bruchstücke einer antiken Schrift über Wetterzeichen', *Sitzungsberichte der philosophisch-historischen Classe der Kaiserlichen Akademie der Wissenschaften* (Wien) 142: 1–41.

West, M.L. (1966) *Hesiod: Theogony*, Oxford: Clarendon Press.

—— (1978) *Hesiod: Works and Days*, Oxford: Clarendon Press.

—— (1983) *The Orphic Poems*, Oxford: Clarendon Press.

—— (1996) 'Hesiod', in Hornblower, S. and Spawforth, A. (eds) *The Oxford Classical Dictionary*, 3rd edn, Oxford: Oxford University Press.

Wieland, W. (1970) *Die aristotelische Physik*, 2nd edn, Göttingen: Vandenhoeck & Ruprecht.

Wildberg, C. (1988) *John Philoponus' Criticism of Aristotle's Theory of Aether*, Berlin/New York: Walter de Gruyter.

Wilkinson, L.P. (1950) 'The intention of Virgil's *Georgics*', *Greece & Rome* 19: 19–28.

—— (1997) *The Georgics of Virgil: A Critical Survey*, Cambridge: Cambridge University Press, 1969; new edn Bristol Classical Press.

Willcock, M.M. (1996) 'Homer', in Hornblower, S. and Spawforth, A. (eds) *The Oxford Classical Dictionary*, 3rd edn, Oxford: Oxford University Press.

Witt, C. (1992) 'Dialectic, motion, and perception: *De Anima* Book I', in Nussbaum, M.C. and Rorty, A.O. (eds) *Essays on Aristotle's* De Anima, Oxford: Clarendon Press.

Woestijne, P. van de (1961) *Le Descriptio orbis terrae d'Avienus: édition critique*, Bruges: De Tempel.

Wöhrle, G. (1985) *Theophrasts Methode in seinen botanischen Schriften*, Amsterdam: B.R. Grüner.

Wood, J.G. (1894) *Theophrastus of Eresus: On Winds and On Weather Signs*, London: Edward Stanford.

Zehnacker, H. (1989) 'D'Aratos à Aviénus: Astronomie et idéologie', *Illinois Classical Studies* 14: 317–29.

Zürcher, J. (1952) *Aristoteles' Werk und Geist*, Paderborn: Verlag Ferdinand Schöningh.

INDEX